The Biomass Assessment Handbook

The Biomass Assessment Handbook

Bioenergy for a Sustainable Environment

Edited by
Frank Rosillo-Calle, Peter de Groot, Sarah L. Hemstock,
and Jeremy Woods

London • Sterling, VA

First published in hardback by Earthscan in the UK and USA in 2007

Reprinted 2007

Paperback edition first published in 2008

ISBN: 978-1-84407-526-3 paperback

Typeset by RefineCatch Limited, Bungay, Suffolk
Printed and bound in UK by TJ International Ltd, Padstow, Cornwall
Cover design by Susanne Harris

For a full list of publications please contact:

Earthscan
8–12 Camden High Street
London, NW1 0JH, UK
Tel: +44 (0)20 7387 8558
Fax: +44 (0)20 7387 8998
Email: earthinfo@earthscan.co.uk
Web: **www.earthscan.co.uk**

22883 Quicksilver Drive, Sterling, VA 20166-2012, US

Earthscan publishes in association with the International Institute for Environment and
Development

A catalogue record for this book is available from the British Library.

Library of Congress Cataloging-in-Publication Data has been applied for

The paper used for the text pages of this book is FSC certified. FSC
(the Forest Stewardship Council) is an international network to
promote responsible management of the world's forests

Printed on totally chlorine-free paper

FSC
Mixed Sources
Product group from well-managed
forests and other controlled sources
Cert no. SGS-COC-2482
www.fsc.org
© 1996 Forest Stewardship Council

Contents

List of Figures, Tables and Boxes

FIGURES

TABLES

Boxes

Editors' Note

The editors have made a substantial contribution to *The Biomass Assessment Handbook*. However, we recognize that since the idea of writing the handbook first started to take shape many people have contributed to it in different ways, directly or indirectly. We would like to acknowledge all of them but they are too many and the time elapsed too long. We have tried to name as many contributors as our records allow, but inevitably some contributors might have been missed to whom we offer our sincere apologies.

Dedication to David Hall

The editors dedicate this manual to the memory of Professor David Oakley Hall, a visionary who tirelessly championed biomass as a modern source of energy and as an agent for social development and environmental protection.

David was an outstanding scientist whose research interests included the bioproductivity of natural grasslands, the production of energy and chemicals from immobilised cells and photosynthesis, earning him the respect of collaborators worldwide. Feeling that the importance of biomass as a modern source of energy and chemicals was seriously overlooked by policymakers in both the developed and the developing world, he took every opportunity to campaign to raise the profile of biomass as a key tool in tackling the major problems facing us today: poverty, the provision of food, energy and shelter for all, and the mitigation of environmental destruction and climate change.

Although a greatly respected academic, David always strived to put theory and knowledge into the real world, encouraging and endorsing practical, multi-disciplinary biomass projects, often in remote locations, that could improve livelihoods and maintain the environment.

David was well aware that the multitude of systems for measuring the various forms of biomass was a factor hindering the assessment of biomass resources and their comparison over time. This made it difficult to present a coherent picture to policy makers and planners of the essential role that biomass energy is now playing in the lives of half the population of the world, and crucial part that modern biomass energy could play in reducing dependence on fossil fuels, mitigating climate change and creating employment. As one attempt to redress this problem, David helped coordinate the multiple inputs from numerous contributors to the first incarnation of this manual. It is sobering to reflect that this was over twenty years ago, since when the problems that David was so acutely aware of have if anything got worse, although the fact that biomass is now more in the headlines would no doubt have raised a wry smile.

David died on 22 August 1999 when he was only 63. His encyclopaedic knowledge, capacity to make things happen and his enthusiasm and energy are sorely missed. All the editors were fortunate enough to have worked with him and benefited from his wisdom, generosity and flare for life. We hope that this manual plays some small part in realising his dream of placing biomass at the heart of human development and environmental welfare.

The Editors

Acknowledgements

The editors would like to thank the following institutions and individuals for their various contributions (financial or otherwise) to this handbook:

AFREPREN*, Al Binger, Biomass Users Network*, Charles Wereko-Brobby, Commonwealth Science Council*, David Hall, FAO*, Gerald Leach, John Soussan, Keith Openshaw, Phil O'Keefe, The Rockefeller Foundation*, Taba Tietema, Woraphat Arthauyti, Gustavo Best, Miguel Trossero, Jonathan M. Scarlock, Alofa Tuvalu Foundation, Gilliane Le Gallic, Fanny Heros, and John Hensford.

* Institutions which provided financial support of various kinds during the lifetime of this project.

Preface

Just as the 20th century saw the rise of oil as a major fuel, the 21st century should be the forum for the emergence of a new mixture of energy carriers increasingly dominated by renewable energy sources and among these, leading the way, bioenergy.

This publication therefore comes at a critical and timely moment.

Traditional bioenergy in the form of fuelwood, charcoal and residues has been with humanity since the discovery of fire, but only in the past 100 years or so has it reappeared in a more advanced and modern version. Let us not forget that biomass gasifiers were used as fuel for vehicles for transport during war times in the 1920s and 1930s and that peanut oil was first used as a fuel in a car engine in 1893 by Rudolf Diesel. Even though bioenergy technological applications were being developed during those times, cheap and ample reserves of fossil fuels in the form of oil and natural gas came into the picture and for over 80 years pushed biofuels to the back seat of energy development.

Things have moved on considerably since then and we are now in a new era where fossil fuels are seeing their last decades of supremacy, where environmental and climate change issues are high on the international agenda and where, for rather unfortunate reasons, energy security has come onto the scene as a major driver of change. These reasons alone would make bioenergy needed as never before. Yet, there are other even more compelling reasons calling for modern bioenergy:

> *the 1.2 billion rural people in the world who are still energy poor and require more affordable, clean and sustainable energy.*

Bioenergy is a locally available energy source with the highest versatility among the renewable energies; that is to say, it can be made available in solid, liquid or gaseous forms. No other energy source can open such new opportunities for agricultural and forest development, additional jobs and enhanced rural infrastructure. It is not by chance that bioenergy development is gaining momentum in many countries, both developing and industrialized, south and north, east to west. In some ways this development seems to be going faster than the advancement of the frontier of knowledge which should guide decisions, set the environmental boundaries, ensure sustainability and assure that social equity accompanies this new field.

That is why this book is so timely – it sheds light and understanding on the key element guiding bioenergy's future.

What is the potential of bioenergy in a particular place and moment?

Without ways of knowing this critical information, decisions can be risky, leading to potential economic losses, environmental disasters and weakened energy scenarios.

The authors of this book are world renowned experts with many years of experience in the forefront of energy and agriculture development, and with internationally recognized professionalism. As committed researchers, they search for better ways and valuable tools to make the road towards sustainable bioenergy systems viable.

The material in this book will certainly be useful to many people in different sectors interested in a better future for society and active in energy development, the future of agriculture, rural populations' incorporation into modern society, an enhanced environment, the mitigation of climate change and, in particular, enhanced food security and stable and fair commodity prices.

I would like to highlight the importance of this book to my organization, the Food and Agriculture Organization of the United Nations (FAO). It will significantly add to the International Bioenergy Platform (IBEP), facilitated by FAO, designed to promote informed policy and technical decision making on bioenergy matters at the national level and to gear international cooperation towards the expected transition to bioenergy in this 21st century.

We therefore very sincerely thank the authors of this book since it is certainly a key element and a major contribution to the IBEP and other bioenergy initiatives worldwide.

Remembering that one definition of information is:

knowledge that causes change

I am pleased and honoured to introduce this book to those in search of information on bioenergy.

Gustavo Best
Senior Energy Coordinator
FAO
May 2006

List of Acronyms and Abbreviations

ABI	Austrian Biodiesel Institute
AMI	American Methanol Institute
BECS	bioenergy with carbon storage
BEDP	bagasse energy development programme
Bl	billion litres
boe	barrel oil equivalent = 42 gallons (US) (1 US gallon = 4.55 litres)
CAI	Current Annual Increment
CCS	Carbon capture, and sequestration
CDM	Clean Development Mechanism
CEB	Central Electricity Board
CHP	combined heat and power
CRF	capital recovery factor
CRI	crop residue index
CT	carbon trading
DBF	densified biomass fuels
DBH	diameter at breast height
DLG	developing countries (also Less Developed Countries)
EEZ	Exclusive Economic Zone
EREC	European Renewable Energy Council
EU	European Union
FAO	Food and Agriculture Organization
FAOSTAT	FAO Statistics Division
FBC	Fluidized bed combustion
FxBC	fixed-bed combustion
G8	Group of Eight
GCV	gross calorific value
GHG	greenhouse gas
GHV	gross heating value
GIS	geographic information system
GPS	Global Positioning System
CNG	compressed natural gas
GTCC	gas turbine/steam turbine combined cycle
HHV	higher heating value
HVS	homogenous vegetation strata
IBEP	International Bioenergy Platform
IBGT	integrated biomass gasifier/gas turbine

IC	internal combustion
IEA	International Energy Agency
IPCC	Intergovernmental Panel on Climate Change
LCA	Life-cycle analysis
LEI	low energy input
LGP	liquid gas petroleum
LHV	lower heating value
LIDAP	light detection and ranging
LPG	liquid petroleum gas
LSP	large scale photography
IEA	International Energy Agency
MAI	mean annual increment
mc	moisture content
MSW	municipal solid waste
Mtoe	million tonnes oil equivalent
NDVI	normalized difference vegetation index
NGO	non-governmental organization
NHV	net heating value
NIR	near infrared
NPP	net primary production (tonnes per hectare per year)
NUE	Nutrient Use Efficiency
ODT	Oven dry ton
ODW	Oven dry weight
OECD	Organisation for Economic Co-operation and Development
PAR	Photosynthetically Active-Radiation
PET	Potential Evapo-Transpiration
PJ	Petajoules
PPA	Power Purchase Agreement
RADAR	radio detection and ranging
RD&D	Research Development and Demonstration
RE	renewable energy
RET	Renewable Energy Technology
SFC	specific fuel consumption
SPOT	Satellites Por l'Observation de la Terre (Earth-observing satellites)
SRC	Short Rotation Coppice
SRES	Special Report on Emissions Scenarios (IPCC)
SST	statistical sampling technique
toe	tonnes oil equivalent = 42 Giga Joules (Mtoe = 1 million toe)
USDOE	USA Department of Energy
UFO	used frying oil
UV	ultraviolet
VME	Vegetable Methyl Ester

Introduction

The idea of writing this handbook is two decades old; many things have changed since it was first conceived as a project of the Commonwealth Science Council. We are now living in a particularly important and exciting moment in the history of bioenergy. In the industrial countries, fossil fuels, especially oil, have for decades dominated the energy we consume, and our entire economy. Transportation systems have become dangerously and overwhelmingly dependent on oil. With current high oil prices, and the realization that this likely to be the dominant feature in the future, we may now be sowing the seeds of a truly fundamental change in the way we use (and produce) energy. In this new scenario, uncertain as it may be, bioenergy is sure to have an increasingly important role.

Bioenergy is no longer a 'transitional energy source' as often portrayed in the past. In fact, many countries around the world have introduced policies in support of bioenergy in the past few years. Further, bioenergy can no longer be passed off as 'the poor person's fuel' but is now recognized as an energy source that can provide the modern consumer with convenient, reliable and affordable services. The focus, therefore, should now be on the development and production of bioenergy for modern applications: a far cry from burning wood on a three stone fire.

However, as the late Anil Agarwal put it, in reality, for most people in the developing world 'life is a struggle for biomass'. For the three-quarters of the world's population who live in developing countries biomass energy is their number one source of primary energy. Some countries, for example Burundi, Ethiopia, Nepal, Rwanda, Sudan and Tanzania, obtain over 95 per cent of their primary energy from biomass. Biomass is not only used for cooking in households and many institutions and service industries, but also for agricultural processing and in the manufacture of bricks, tiles, cement, fertilizers, etc. There is often a substantial use of biomass for these non-cooking purposes, especially in and around towns and cities.

For the majority of the world's people, biomass will continue to be the prime source of energy for the foreseeable future. A study by the IEA (2002) states that:

> Over 2.6 billion people in developing countries will continue to rely on biomass for cooking and heating in 2030 . . . an increase of more than 240 million. [In 2030] biomass use will still represent over half of residential energy consumption.

For the majority of the world's people, therefore, biomass will continue to be the prime source of energy for the foreseeable future.

TRADITIONAL AND MODERN USES OF BIOMASS ENERGY

Biomass energy is readily obtained from wood, twigs, straw, dung, agricultural residues, etc. Biomass is burnt either directly for heat, or to generate electricity, or can be fermented to alcohol fuels, anaerobically digested to biogas, or gasified to produce high-energy gas.

Along with agricultural expansion to meet the food needs of increasing populations, the overuse of traditional biomass energy resources has led to the increasing scarcity of hand-gathered fuelwood, and to the problems of deforestation and desertification. Even so, there is an enormous untapped biomass potential, particularly in the improved utilization of existing forest and other land resources and in higher plant productivity.

ENERGY EFFICIENCY

Biomass in its 'raw' form is often burnt very inefficiently, so that much of the energy is wasted. For example, where fuelwood is used for cooking in rural areas, the per capita use of energy is several times greater than when gaseous or liquid fuels are used, and is comparable with per capita energy use in automobiles in western Europe in the mid-1970s!

Energy planners should consider the application of advanced technologies to convert raw biomass into modern, convenient energy carriers such as electricity, liquid or gaseous fuels, or processed solid fuels, in order to increase the energy that can be extracted from biomass.

MULTIPLE USES

In addition to food and energy, biomass is a primary source of many essential daily materials. The multiple uses for biomass are summarized as the six 'fs': food, feed, fuel, feedstock, fibre and fertilizer. Biomass products are frequently a source of a seventh 'f' – finance.

Because of the almost universal, multipurpose dependence on biomass, it is important to understand interrelations between these many uses, and to determine the possibilities for more efficient production and wider uses in future. The success of any new form of biomass energy probably depends on the use of reasonably advanced technology. It is therefore important to investigate the potential for biomass in the form of modern biofuels powering gas turbines and as alcohol fuels, for example.

LAND MANAGEMENT

The optimum sustainable production and use of biomass is, ultimately, really a problem of land management. New and traditional agricultural agro-forestry and intercropping systems maximize energy and food production. At the same time these systems diversify land use by producing ancillary benefits such as fodder, fertilizers, construction materials and medicines, etc. while at the same time increasing environmental protection by, for example, maintaining soil fertility and structure under long-term cultivation.

Large-scale energy plantations require also that a country or region has a policy on how to use its land. Such a policy would go far beyond energy to include policies on food production and prices, land reform, food exports and imports, tourism and the environment. Lack of such integrated policies often gives rise to conflicts between land use for fuel, food and for other uses, which can result in the indiscriminate clearing of forests and savannah for agricultural expansion.

ENVIRONMENTAL FACTORS

The sustainable production and conversion of plants and plant residues into fuels offers a significant opportunity for alleviating the pressure on forests and woodlands for use as fuel. Along with agricultural clearances, these pressures have been the major threats to forest and tree resources, wetlands, watersheds and upland ecosystems.

Biomass fuels also have an important role to play in the mitigation of climate change. Using modern energy conversion technologies it is possible to displace fossil fuels with an equivalent biofuel. When biomass is grown sustainably for energy, with the amount grown equal to that burned for a given period, there is no net build-up of CO_2, because the amount of CO_2 released in combustion is compensated for by that absorbed by the growing energy crop.

Ever-pressing environmental concerns over devegetation and deforestation, desertification, and the role of CO_2 in climate change emphasize it is crucial that we move to the sustainable and efficient use of biomass energy both as a traditional fuel and in modern, greenhouse gas-neutral commercial applications.

WHY THIS HANDBOOK?

Despite the overriding importance of biomass energy, its role continues to be largely unrecognized. Many developing countries are currently experiencing acute shortages of biomass energy. Programmes to tackle this breakdown in the biomass system will require detailed information on the consumption and supply of biomass. Governments and other agencies need detailed knowledge of energy demand along with information on the annual yield and growing stock of biomass resources in order to plan for the future. Clearly some standardized measurements are required of the supply and demand for the various forms of biomass energy, similar to those available for fossil fuels.

However, there is a general lack of information on the requirements and use of biomass energy. Data are often inaccurate. Furthermore, there are no standard measuring and accounting procedures for biomass, so it is often impossible to make comparisons between sets of existing data. This handbook is intended to provide a practical, common methodology for measuring and recording the consumption and supply of biomass energy. The book is designed to appeal to a wider audience and thus we have tried to avoid including too many technical details.

As indicated in Chapter 1, biomass resources are potentially the world's largest and most sustainable source of energy, which in theory at least could contribute to over 800 exajoules (EJ), without affecting the world's food supply. By comparison, current energy use is just over 400 EJ. It is not surprising that biomass features strongly in virtually all the major global energy supply scenarios.

This handbook is the result of many years of personal experience. We have gathered material from own fieldwork, teaching courses and other existing materials. We wanted to demonstrate the importance of bioenergy and show how to assess such an important resource step by step. This handbook puts particular emphasis on traditional bioenergy applications, but modern uses are also considered. This is because the traditional applications represent the most serious difficulties when it comes to measurement. It is important to note that measuring techniques, as with everything else, are not static but constantly changing. This handbook is intended to help those people interested in understanding the biomass resource base and the techniques to measure it; this is described in the various chapters as follows. Conversion technologies are not described in any detail in this volume since they will be dealt with in a second, forthcoming volume.

Chapter 1 presents a short overview of the biomass resource potential, current and future uses (various scenarios), technology options, traditional versus modern applications and difficulties with quality data. The aim is to familiarize the non-specialist reader with bioenergy, the importance of biomass as an energy source and its likely energy contribution to the world energy matrix in the future.

Chapter 2 deals with the problems in measuring biomass, biomass classification, general methods for assessing biomass resources and land use assessment. It also looks briefly at remote sensing, biomass flow charts, units for measuring biomass (e.g. stocking, moisture content and heating values), weight versus volume, calculating energy values, and finally considers possible future trends.

Chapter 3 brings together various types of biomass (woody and herbaceous), planning issues (land constraints, land uses, tenure rights), climatic issues, etc. It looks in some detail at the most important methods for accurately measuring the supply of woody biomass for energy, and in particular techniques for forest mensuration, determining the weight and volume of trees, measuring the growing stock and yield of trees, and measuring the height and bark, the energy available from dedicated energy plantations, agro-industrial plantations and processed woody biomass (woody residues, charcoal).

Chapter 4 is concerned primarily with traditional agricultural crops that offer potential as a source of energy. New crops currently being considered as possible dedicated energy crops, for example, miscanthus, reed canary grass and switchgrass, are included in the analysis. It also examines densified biomass (briquettes, pellets, wood chips), which are increasingly being traded internationally; looks in some detail at secondary fuels (biodiesel, biogas, ethanol, methanol and hydrogen); and briefly reviews tertiary fuels (municipal solid wastes (MSW)) as their development can have major impacts on biomass resources. Finally, the chapter assesses animal traction, which still plays a major role in many countries around the world and also has an impact on biomass resources.

The prime emphasis of Chapter 5 is the assessment of biomass supply and consumption patterns. It looks at designing biomass energy consumption surveys, questionnaire design, implementing surveys and different patterns of biomass energy consumption in the domestic sector, fuel uses and finally provides various examples of primary energy consumption in small islands.

Chapter 6 assesses the main techniques used in remote sensing for measuring biomass, particularly the estimation of woody biomass production for use as a source of energy. It assesses the utility of the techniques for estimating biomass at a project or a large plantation or at a landscape level, and not for national or global forest biomass assessment. This chapter presents bioenergy utility and forest or plantation managers with various remote sensing techniques for estimating, monitoring or verifying biomass production or growth rates. Further, these techniques can be used for estimating carbon stock changes in carbon sequestration projects.

Satellite imagery-based remote sensing techniques provide an alternative to traditional methods for estimating or monitoring or verifying an area of forest or plantation, and biomass production or growth rates. Remote sensing techniques provide spatially explicit information and enable repeated monitoring, even in remote locations, in a cost-effective way.

Chapter 7 comprises five case studies, each of which deals in detail with a particular aspect of bioenergy. They are used to illustrate step-by-step methods for calculating biomass resources and uses in modern applications, based on fieldwork experiences, or to illustrate potentially major changes or trends in a particular area. These are:

1 biotrade, which examines the development of international biotrade in bio-energy and its wider implications;
2 biogas use in small island communities;
3 the utilization of biodiesel from coconuts in small islands (e.g. Tuvalu);
4 how to build a modern biomass market (e.g. Austria); and
5 the last case study looks at the potential impacts of biomass energy, carbon sequestration and climate change.

Finally, various Appendices have been added to assist the reader in pursuing some technical data in greater detail.

Reference

IEA (2002) *Energy Outlook 2000–2030*, IEA, Paris, www.iea.org

Overview of Bioenergy

Frank Rosillo-Calle

INTRODUCTION

Bioenergy is not an energy source in transition, as it is often portrayed, but a resource that is becoming increasingly important as a modern energy carrier. This chapter provides an overview of bioenergy and its potential as the world's largest source of renewable energy. It examines the role of biomass energy, its potential (traditional versus modern applications and linkages between the two), details the difficulties in compiling information and classifying biomass energy, looks at the barriers to the use of biomass energy and, finally, examines the possible future roles of bioenergy. It is not possible to present a detailed analysis and thus it is strongly recommended that the reader consults the reference section at the end of the chapter.

HISTORICAL ROLE OF BIOENERGY

Throughout human history biomass in all its forms has been the most important source of all our basic needs, often summarized as the six 'fs': food, feed, fuel, feedstock, fibre and fertilizer. Biomass products are also frequently a source of a seventh 'f' – finance. Until the early 19th century biomass was the main source of energy for industrial countries and, indeed, still continues to provide the bulk of energy for many developing countries.

Past civilizations are the best witnesses to the role of bioenergy. Forests have had a decisive influence on world civilizations, which flourished as long as towns and cities were backed up by forests and food producing areas. Wood was the foundation on which past societies stood. Without this resource, civilization failed – forests were for them what oil is for us today (Rosillo-Calle and Hall, 2002; Hall et al, 1994). For example, the Romans used enormous quantities of wood for building, heating and for all sorts of industries. The Romans commissioned ships to bring wood from as far away as France, North Africa and Spain. Material for architecture and shipbuilding, fuels for metallurgy, cooking, cremation and heating etc. left Crete, Cyprus, Mycenaean Greece and many areas around Rome bereft of much of their forests (Perlin and Jordan, 1983). When forests were exhausted, these civilizations began to decline.

The first steps towards industrialization were also based on biomass resources. Take charcoal, for example, used in iron smelting for thousands of years. Archaeologists have suggested that charcoal-based iron making was responsible for large-scale deforestation near Lake Victoria, Central Africa, about 2500 years ago. In modern times, Addis Ababa is a good example of dependency on fuelwood. Ethiopia did not have a modern capital until the establishment of modern eucalyptus plantations, which, early in the 20th century, allowed the government to remain in Addis Ababa on a continuous basis. Before a sustainable source of biomass was secured, the government was forced to move from region to region as the resources became exhausted (Hall and Overend, 1987).

In fact, some historians have argued that the United States and Europe would not have developed without abundant wood supplies, since the Industrial Revolution was initially only possible due to availability of biomass resources. Britain is an excellent example of a country which was able to become one of the world's most powerful nations thanks largely to its forests. Initially, forests, mainly of oak, covered two-thirds of Britain. The wood and charcoal produced from these forests was the basis for the Industrial Revolution, and continued to fuel industrial development in Britain until well into the 19th century (Schubert, 1957).

Worldwide, biomass fuels are used for cooking in households and many institutions and cottage industries, ranging from brick and tile making, metalworking, bakeries, food processing, weaving, restaurants and so forth. More recently, many new plants are being set up to provide energy from biomass directly through combustion, to generate electricity, or in combined heat and power (CHP) facilities or ethanol via formentation. Contrary to the general view, biomass utilization worldwide remains steady or is growing for three broad reasons:

- population growth;
- urbanization and improvement in living standards;
- increasing environmental concerns.

THE CURRENT ROLE OF BIOENERGY

Today, biomass energy continues to be the main source of energy in many developing nations, particularly in its traditional forms, providing on average

35 per cent of the energy needs of three-quarters of the world's population. This rises to between 60 and 90 per cent in the poorest developing countries. However, modern biomass energy applications are increasing rapidly both in the industrial and developing countries, so that they now account for 20–25 per cent of total biomass energy use. For example, the United States obtains about 4 per cent and Finland and Sweden 20 per cent of their primary energy from biomass.

Biomass energy is not a transition fuel as it has often been portrayed, but a fuel that will continue to be the prime source of energy for many people for the foreseeable future. For example, an IEA (2002) study concluded:

> *Over 2.6 billion people in developing countries will continue to rely on biomass for cooking and heating in 2030 . . . this is an increase of more than 240 million from current use. In 2030 biomass use will still represent over half of residential energy consumption . . .*

Because of the almost universal, multipurpose dependence on biomass, it is important to understand the interrelations between these many uses, and to determine the possibilities for more efficient production and wider uses in future. The success of any new form of biomass energy will most probably depend upon the use of reasonably advanced technology. Indeed, if bioenergy is to have a long-term future, it must be able to provide what people want: affordable, clean and efficient energy forms such as electricity, and liquid and gaseous fuels. This also entails direct competition with other energy sources.

Table 1.1 *Scenarios of potential biomass contribution to global primary energy (EJ)*

Scenario	Biomass primary energy supply		
	2025	2050	2100
Lashof and Tirpack (1991)[a]	130	215	
Greenpeace (1993)[a]	114	181	
Johansson et al (1993)[a]	145	206	
WEC (1994)[a]	59	94–157	132–215
Shell (1996)	85	200–220	
IPCC (1996) – SAR	72	280	320
IEA (1998)	60		
IIASA/WEC (1998)	59–82	97–153	245–316
IPCC (2001) – TAR	2–90	52–193	67–376

Note: Present biomass energy use is about 55 EJ/year.
Sources:
[a] See Hall et al, 2000 for further details.
'TAR' – IPCC Third Assessment Report, 2001.
'SAR' – IPCC Second Assessment Report, 1996.

BIOMASS POTENTIAL

Biomass features strongly in virtually all the major global energy supply scenarios, as biomass resources are potentially the world largest and most sustainable energy source. Biomass is potentially an infinitely renewable resource comprising 220 oven dry tonnes (odt), or about 4500 exajoules (EJ), of annual primary production; the annual bioenergy potential is about 2900 EJ (approximately 1700 EJ from forests, 850 EJ from grasslands and 350 EJ from agricultural areas) (Hall and Rao, 1999). In theory, at least, energy farming in current agricultural land alone could contribute over 800 EJ without affecting the world's food supply (Faaij et al, 2002).

There are large variations between the many attempts to quantify the potential for bioenergy. This is due to the complex nature of biomass production and use, including such factors as the difficulties in estimating resource availability, long-term sustainable productivity and the economics of production and use, given the large range of conversion technologies, as well as ecological, social, cultural and environmental considerations. Estimating biomass energy use is also problematic due to the range of biomass energy end-uses and supply chains and the competing uses of biomass resources. There is also considerable uncertainty surrounding estimates of the potential role of dedicated energy forestry/crops, since the traditional sources of biomass they could replace, such as residues from agriculture, forestry and other sources have a much lower and varied energy value. Furthermore, the availability of energy sources, including biomass, varies greatly according to the level of socio-economic development. All these factors make it very difficult to extrapolate bioenergy potential, particularly at a global scale.

All major energy scenarios include bioenergy as a major energy source in the future, as illustrated in Table 1.1. For the reasons given above, there are very large differences in these estimates, so these figures should be considered only as rough estimates. The figures in Table 1.1 are based on estimates of future global energy needs and the determination of the related primary energy mix, including biomass energy share, based on resource, cost and environmental constraints (i.e. a 'top-down' approach). In order to achieve realistic scenarios for biomass energy use and its role in satisfying future energy demand and environmental constraints, it is important to reconcile 'top-down' and 'bottom-up' modelling approaches.

Faaij et al (2002) have also developed a scenario for the global potential of bioenergy by 2050, summarized in Table 1.2. This potential ranges from 40 to 1100 EJ – the most pessimistic scenario, with no land available for energy farming, and thus bioenergy will come only from residues – to 200 to 700 EJ for most optimistic scenario, which considers that intensive agriculture takes place only in the better quality soils, while land of lower quality will be used for energy forestry/crops.

Table 1.2 *Overview of the global potential bioenergy supply in the long term for a number of categories and the main preconditions and assumptions that determine these potentials*

Biomass category	Main assumptions and remarks	Potential bioenergy supply in 2050
Category I: Energy farming on current agricultural land	Potential land surplus: 0–4 global hectares (gha) (more average: 1–2 gha). A large surplus requires structural adaptation of high energy input (HEI) agricultural production systems. When this is not feasible, the bioenergy potential could be reduced to zero as well. On average higher yields are likely because of better soil quality: 8–12 dry tonne/ha*year is assumed (*)	0–870 EJ (more average development: 140–430 EJ)
Category II: Biomass production on marginal lands	On a global scale a maximum land surface of 1.7 gha could be involved. Low productivity of 2–5 dry tonne/ha/year. (*heating value: 19 GJ/tonne dry matter) The supply could be low or zero due to poor economics or competition with food production	(0) 60–150 EJ
Category III: Bio-materials	Range of the land area required to meet the additional global demand for biomaterials: 0.2–0.8 gha (average productivity: 5 dry tonnes/ha/year*). This demand should come from category I and II if the world's forests are unable to meet the additional demand. If they are, however, the claim on (agricultural) land could be zero	*Minus* (0) 40–150 EJ
Category IV: Residues from agriculture	Estimates from various studies. Potential depends on yield/ product ratios and the total agricultural land area as well as type of production system: low energy input (LEI) systems require reuse of residues for maintaining soil fertility. HEI systems allow for higher utilization rates of residues	Approx. 15EJ
Category V: Forest residues	The (sustainable) energy potential of the world's forests is unclear. Part is natural forest (reserves). Range is based on literature data. Low value: figure for sustainable forest management; high value: technical potential	(0) 14–110 EJ
Category VI: Dung	Use of dried dung. Low estimate based on global current use. High estimate: technical potential. Utilization (collection) on longer term is uncertain	(0) 5–55 EJ
Category VII: Organic wastes	Estimate on basis of literature values. Strongly dependent on economic development, consumption and the use of biomaterials. Figures include the organic fraction of MSW and waste wood. Higher values possible by more intensive use of biomaterials	5–50 (+) EJ (**)
Total	Most pessimistic scenario: no land available for energy farming; only utilization of residues. Most optimistic scenario: intensive agriculture concentrated on the better quality soils (between brackets: more average potential in a world aiming for large scale utilization of bioenergy)	40–1100 EJ (200–700 EJ)

Notes:
(*) Heating value: 19 GG/tonne dry matter.
(**) The energy supply of biomaterials ending up as waste can vary between 20–55 EJ (or 1100–2900 Mtonne dry matter per year. This range excludes cascading and does not take into account the time delay between production of the material and 'release' as (organic) waste.
Source: Faaij et al (2002).

Appendix 1.1 shows the role of bioenergy in 2000 by major regions. It is important to state that the figures for biomass energy should be regarded as conservative as most official agencies tend to downgrade bioenergy. Appendix 1.2 also shows the current and estimated global energy consumption from 2001 to 2040, based on estimates of the European Renewable Energy Council (EREC), with particular emphasis on renewable energy (RE), which indicates a tenfold increase over this period. This increase is, however, very unevenly distributed around the world, with some countries making little use of RE and others (e.g. those in the European Union (EU)) experiencing a significant increase in RE based on current favourable policies. This unevenness is also common within countries where there are strong regional variations. For example, in Navarra, a region located in northern Spain, at the end of 2005 about 95 per cent of electricity was generated from RE, compared with less than 10 per cent in other regions of the country. The EREC's scenarios do not take into account technological developments in other energy sectors (e.g. oil and gas), which could have a major impact on future development of RE in general and bioenergy in particular.[1]

Energy demand (in particular primary commercial energy consumption) has increased rapidly at an average rate of 2 per cent per year. This growth is particularly acute in some developing countries (DLG) such as China and India, where energy demand increased by nearly 32 per cent and 46 per cent respectively in the period 1993–2002, compared with a world average of 14.2 per cent over the same period. Many other DLGs have also shown a rapid increase in demand for energy (e.g. 37.3 per cent in DLG Asia-Pacific countries; 28.5 per cent in Africa and 21.6 per cent in Central and South America) over the same period (Bhattacharya, 2004). Meeting such growing energy demand is posing serious energy supply problems.

In many industrial countries significant areas of cropland are becoming available in response to pressure to reduce agricultural surpluses, particularly in the EU and the United States. For these countries there is a pressing need to find alternative economic opportunities for the land and associated rural populations, and biomass energy systems could provide such opportunities in some, although not all, these countries. An important strategic element in developing a biomass energy industry is the need to address the introduction of suitable crops, logistics and conversion technologies. This may involve a transition over time to more efficient crops and conversion technologies.

Residues versus energy plantations

Residues (all sources) are currently at the heart of bioenergy. Residues are a large and underexploited potential energy resource, almost always underestimated, and represent many opportunities for better utilization. There have been many attempts to calculate this potential but it is quite difficult for the reasons discussed above.

Table 1.3 shows that the energy potentially available from crop, forestry and animal residues globally is about 70 EJ. However, there is considerable variation in the estimates, which vary from around 3.5 to 4.2 Gt/yr for agriculture and 800–900 Mt/yr for forestry residues, and thus should be regarded as rough indications only. The estimates given in Table 1.3 are based on the energy content of potentially harvestable residues based on residue production coefficients applied to FAOSTAT data on primary crop and animal production. Forestry residues are calculated from FAOSTAT 'Roundwood' and 'Fuelwood and Charcoal' production data, again using standard residue production coefficients (see Woods and Hall, 1994 for more details).

In the forestry sector, and particularly commercial forestry, most of residues are used to generate its own energy. This practice of using forest residues to provide power for the industry is increasing around the world. On the other hand, in countries such as China currently undergoing a rapid fuel switch, some agricultural residues such as straw are left to rot in the fields or simply burned, while in other countries such as the United Kingdom it is utilized in modern combustion plants. Globally, about 50 per cent of the potentially available residues are associated with the forestry and wood processing industries; about 40 per cent are agricultural residues (e.g. straw, sugarcane residues, rice husks and cotton residues) and about 10 per cent animal manure.

One important rationale for the development of residue-based bioenergy industries is that the feedstock costs are often low or even negative where a 'tipping fee' is levied. There are a number of important factors that need to be addressed when considering the use of residues for energy. First, there may be other important alternative uses, e.g. animal feed, erosion control, use as animal bedding, as fertilizer (dung), etc. Second, it is not clear just what quantity of residues is available, as there is no agreed common methodology for determining what is and what is not a 'recoverable residue'. Consequently, estimates of residues often vary by a factor of five.

Table 1.3 *Energy potential from residues (EJ)*

	Crop	Forest	Dung	Total
World	24	36	10	70
OECD of which:	7	14	2	24
N. America	4	9	0.7	14
Europe	3	5	1	9
Asia-Pacific/Oceania	0.8	0.8	0.4	2

Note: Rounding errors may mean that columns do not add, see Table 15 for details.
Source: Bauen et al (2004).

Energy forestry crops

The production of energy forestry crops is intensive in its land use requirement. Energy forestry crops can be produced in two main ways:

1   as dedicated plantations in land specifically devoted to this end, and
2   intercropping with non-energy forestry crops.

The future role of energy plantations is difficult to predict, since this will depend on many interrelated factors including land availability, costs and the existence of other alternatives. Estimates of land requirements have ranged from about 100 Mha to over one billion ha, with enormous variations from region to region around the world. Table 1.4 shows an estimate of the potential contribution of bioenergy plantations by 2020.

Table 1.4 *Potential contribution of biomass energy from energy plantations by 2020*

	Potential based on 5% of crop, forest and wood land and average 150 GJ/ha yield (EJ)	Share of primary energy – 1998 (%)	Share of electricity consumption – 1998 (%)	Share of primary energy – 2020 (%)	Share of electricity consumption – 2020 (%)
World	42.5	12	29	6	17
OECD of which:	11.6	5	12	3	9
N. America	7.8	7	16	5	13
Europe	2.2	4	7	2	6
Asia-Pacific/ Oceania	1.5	4	10	5	7

Source: Bauen et al (2004)

Large areas of former cropland and unexploited plantation, forests and woodlands are likely to be available in most countries to provide a significant biomass energy contribution. However, there is considerable uncertainty as to the real potential of this alternative as explained below. The technical potential for energy provision from dedicated 'energy crops' including short, medium and long rotation forestry is indeed large. For example, half the present global area of cropland would be sufficient to satisfy current primary energy needs based on a global average yield of 10 dry tonnes of biomass per hectare. However, there are still many uncertainties.

Overall, it seems that the predictions of very large-scale energy plantation are unlikely to materialize, despite the potential availability of land,[2] for the following reasons:

- Degraded land is less attractive than good quality land due to higher costs and lower productivity, although the importance of bringing degraded land into productive use is recognized.
- Capital and financial constraints, particularly in developing countries.
- Cultural practices, mismanagement, perceived and potential conflict with food production, population growth, etc.
- The need for productivity to increase far beyond what may be realistically possible, although large increases are possible.
- Increasing desertification problems and the impacts of climate change in agriculture which presently are too unpredictable.
- Emergence of other energy alternatives (e.g. clean coal technology, wind power, solar, etc.).
- Water constraints (Rosillo-Calle and Moreira, 2006).

Thus, the fundamental problem is not availability of biomass resources but the sustainable management and the competitive and affordable delivery of modern energy services. This implies that all aspects both production and use of bio-energy must be modernized and, most importantly, maintained on a sustainable and long-term basis.

Biomass fuels also have an increasingly important role to play in the welfare of the global environment. Using modern energy conversion technologies it is possible to displace fossil fuels with an equivalent biofuel. When biomass is grown sustainably for energy there is no net build-up of CO_2, assuming that the amount grown is equal to that burned, as the CO_2 released in combustion is compensated for by that absorbed by the growing energy crop. The sustainable production of biomass is therefore an important practical approach to environmental protection and longer-term issues such as reforestation and revegetation of degraded lands and in mitigating global warming. Bioenergy can play a significant role both as a modern energy source and in abating pollution.

Indeed, a combination of environmental considerations, social factors, the need to find new alternative sources of energy, political necessities and rapidly evolving technologies are opening up new opportunities for meeting the energy needs from bioenergy in an increasingly environment-conscious world. This is reflected in the current worldwide interest in RE in general and bioenergy in particular. Concerns with climate change and environment are playing a significant role in promoting bioenergy, although there is still considerable uncertainty as to what the ultimate effects will be.

CONTINUING DIFFICULTIES WITH DATA/CLASSIFICATION
OF BIOENERGY

Information on the production and use of bioenergy is plagued with difficulties due to lack of reliable long-term data. Even when available, these data are often inaccurate and too site-specific. Biomass energy, particularly in its traditional forms, is difficult to quantify because there are no agreed standard units for measuring and quantifying various forms of biomass, making it difficult to compare data between sites. Furthermore, as traditional biomass is an integral part of the informal economy, in most cases it never enters official statistics. Given the nature of the biomass resource, developing and maintaining a large bioenergy databank is very costly. Biomass is generally regarded as a low status fuel, as the 'poor person's fuel'. Traditional uses of bioenergy, for example fuelwood, charcoal, animal dung and crop residues, are also often associated with increasing scarcity of hand-gathered fuelwood, and have also unjustifiably been linked to deforestation and desertification.

Despite the overwhelming importance of biomass energy in many developing countries, the planning, management, production, distribution and use of biomass, hardly receives adequate attention among policy makers and energy planners. When relevant policy provisions are made, they are hardly ever put into practice due to a combination of factors such as budgetary constraints, lack of human resources, the low priority assigned to biomass, a lack of data, etc. Even today some countries fail to produce adequate and reliable data on bioenergy; or worse, they fail to come up with any meaningful data at all.

Considerably more long-term reliable data on all aspects of biomass production and use are still required. For example, there are very few energy flow charts at national levels apart from those produced for the United States, Zimbabwe and Kenya (Ross and Cobb, 1985; Hemstock and Hall, 1993; Senelwa and Hall, 1993), respectively. Biomass energy flows are important because they are very useful methods of representing data and can provide a good overview of national, regional and local conditions when the base data are available for their compilation.

However, in recent years, thanks to considerable efforts by some international agencies (i.e. FAO, IEA and national governments), data have improved significantly, particularly in industrial countries, although in poorer countries a lack of good biomass data often remains a serious problem. This is particularly the case with economic data, which are not readily available or are quoted in a way that makes comparisons very difficult. The inability to fully address the indigenous biomass resource capability and its likely contribution to energy and development is still a serious constraint to the full realization of this energy potential.

A further constraint is confusion with the terminology. FAO has been trying to address this problem for years and after many consultations a document is currently available (www.fao.org/doccrep/007) that will go some way towards solving some of these problems (see FAO/WE, 2003).

Overall, despite the overriding importance of biomass energy, its role is still not fully recognized. There is surprisingly little reliable and detailed information on the consumption and supply of biomass in many countries, but worse, unstandardized systems for measurements and accounting procedures are still common. This serious lack of information is preventing policy makers and planners from formulating satisfactory sustainable energy policies. Programmes to tackle this breakdown in the biomass system will require detailed information on the consumption and supply of biomass (e.g. annual yield and growing stock of biomass resources) in order to plan for future. Clearly, some standardized measurements are required to put biomass energy on a comparable basis with fossil fuels. Hopefully, this handbook will make some contributions by addressing these issues.

The FAO classifies bioenergy into three main groups:

1   woodfuels
2   agro-fuels, and
3   urban waste-based fuels (see Figure 1.1).

Biomass can also be classified as:

•   traditional bioenergy (firewood, charcoal, residues), and
•   modern biomass (associated with industrial wood residues, energy plant-
    ations, use of bagasse, etc.). See also Appendix I Glossary of Terms.

TRADITIONAL VERSUS MODERN APPLICATIONS OF BIOENERGY

Traditional uses of biomass in its 'raw' form are often very inefficient, wasting much of the energy available, and are also often associated with significant negative environmental impacts. Modern applications are rapidly replacing trad-itional uses, particularly in the industrialized countries. Changes are also occur-ring in many developing countries, although very unevenly. For example, in China the use of bioenergy is declining rapidly while in India it is increasing. However, in absolute terms the use of traditional bioenergy continues to grow due to rapid population increases in many developing countries, increasing demand for energy and a lack of accessible or affordable alternative energy sources, especially for the urban poor and a large proportion of the population living in the rural areas, etc.[3]

Modern applications require capital, skills, technology, market structure and a certain level of development, all of which are lacking in most of the rural areas of DLGs.

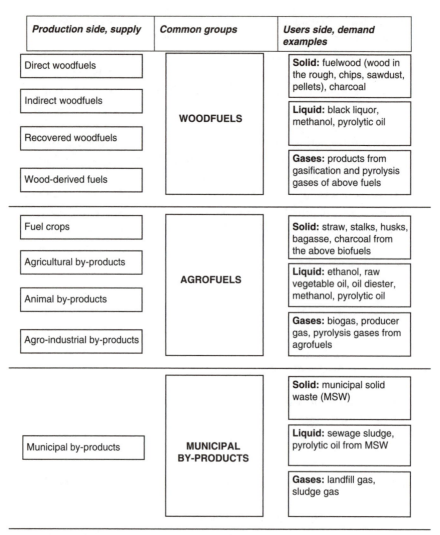

Production side, supply	Common groups	Users side, demand examples
Direct woodfuels	**WOODFUELS**	**Solid:** fuelwood (wood in the rough, chips, sawdust, pellets), charcoal
Indirect woodfuels		**Liquid:** black liquor, methanol, pyrolytic oil
Recovered woodfuels		**Gases:** products from gasification and pyrolysis gases of above fuels
Wood-derived fuels		
Fuel crops	**AGROFUELS**	**Solid:** straw, stalks, husks, bagasse, charcoal from the above biofuels
Agricultural by-products		**Liquid:** ethanol, raw vegetable oil, oil diester, methanol, pyrolytic oil
Animal by-products		**Gases:** biogas, producer gas, pyrolysis gases from agrofuels
Agro-industrial by-products		
Municipal by-products	**MUNICIPAL BY-PRODUCTS**	**Solid:** municipal solid waste (MSW)
		Liquid: sewage sludge, pyrolytic oil from MSW
		Gases: landfill gas, sludge gas

Note: For further details visit www.fao.org/doccrep/007/ (Unified Bioenergy Terminology).

Figure 1.1 *Biofuel classification scheme*

Traditional biomass use

Traditional uses of biomass have been estimated at between 700 Mtoe to 1200 Mtoe, depending on the source. These are rough estimates since, as already mentioned, traditional uses are at the core of the informal economy and never enter the official statistics. Traditional sources of bioenergy in the poorest DLGs still represent the bulk of biomass energy use. For example, in Burundi, Ethiopia, Mozambique, Nepal, Rwanda, Sudan, Tanzania and Uganda, about 80–90 per cent of energy comes from biomass.

The efficiency in the use of traditional bioenergy in DLGs varies considerably from between about 2 per cent to 20 per cent. This compares to 65–80 per cent (or even 90 per cent) using modern technologies, particularly in industrial countries such as Austria, Finland and Sweden (Rosillo-Calle, 2006).

Table 1.5 shows traditional bioenergy consumption in selected industrial and DLGs between 1980 and 1997. The general trend is for an increase in consumption in industrial countries and a decrease in DLGs. These figures have to be put into context. First, in the industrial countries the increases are mostly the result of deliberate policies to support renewable energies and also reflect a lower growth in energy demand. In the case of DLGs these increases reflect rapid energy demand at all levels triggered by economic development and low energy consumption in comparison to industrial countries and the desire to move away from traditional bioenergy. Table 1.6 illustrates the considerable potential for improving energy efficiency, which could be achieved with better management practices and minor technological improvements.

A study by Bhattacharya et al (1999) shows that there is an enormous potential to improve energy efficiency just through small improvements. The authors identified a potential saving of 328 Mt of biomass fuels in the eight

Table 1.5 *Traditional bioenergy consumption in selected countries*

	Percentage of traditional biomass in 1980[a]	Percentage of traditional biomass in 1997[a]
Country (DVD)		
Denmark	0.4	5.9
Japan	0.1	1.6
Germany	0.3	1.3
Netherlands	0.0	1.1
Sweden	7.7	18.0
Switzerland	0.9	6.0
USA	1.3	3.8
Country (DLG)		
Brazil	35.5	28.7
China	8.4	5.7
India	31.5	30.7
Malaysia	15.7	5.5
Nicaragua	49.2	42.2
Peru	15.2	24.6
Philippines	37.0	26.9
Sri Lanka	53.5	46.5
Sudan	86.9	75.1
Tanzania	92.0	91.4
Thailand	40.3	24.6

Note:
[a] Biomass energy consumption as percentage of total energy use.
Source: Bhattacharya (2004) (tables 2 and 3).

Table 1.6 *Potential savings from bioenergy through efficiency improvement, selected Asian developing countries (Mt/yr, based on the stated year)*

Country	Year	Fuelwood	Agricultural residues	Animal dung	Charcoal
China	1993	51.6	77.2	2.9	–
India	1991	69.5	20.8	32.3	0.5
Nepal	1993	3.1	1.2	0.8	–
Pakistan	1991	17.5	7.3	8.3	–
Philippines	1995	7.6	2.3	–	0.3
Sri Lanka	1993	2.6	0.5	–	–
Vietnam	1991	15.8	3.9		0.1
Total		**167.7**	**113.2**	**44.3**	**0.9**

Source: Bhattacharya et al (1999); Bhattacharya (2004).

selected Asian countries alone. In addition, replacing all traditional stoves with improved stoves in those countries will save a further 296 Mt annually. Savings in firewood in the domestic sector were estimated at about 152 Mt or 43 per cent of the fuelwood used in households (Bhattacharya, 2004). These figures give an indication of the enormous inefficiency in traditional bioenergy applications and the urgent need for improved technologies.

Modern applications

As was clearly reflected in the 2004 Bonn Conference,[4] which was attended by representatives from 154 countries, concerted support for RE is leading to a rapid, albeit varying, increase in modern applications of bioenergy around the world. The modernization of biomass embraces a range of technologies that include combustion, gasification and pyrolysis for:

- household applications, e.g. improved cooking stoves, use of biogas, ethanol, etc.
- small cottage industrial applications, e.g. brick-making, bakeries, ceramics, tobacco curing, etc.
- large industrial applications, e.g. CHP, electricity generation, etc. (Rosillo-Calle, 2006).

The current installed biomass-based electricity generation is over 40 GW worldwide, but growing rapidly. For example, China is expected to have between 3.5 and 4.1 GW and India from 1.4 to 1.7 GW by 2015 compared to less than 200 MW today (Bhattacharya, 2004) (see the section on 'Technology options' below). Transport is one of the most rapidly increasing applications using, for example, ethanol and biodiesel, which are the best alternatives to petrol and

Table 1.7 *World ethanol production, 2000–2004 (billion litres)*[a]

	2004	2003	2002	2001	2000
Europe	4.01	4.27	4.08	4.03	3.56
EU-15	2.58	2.37	2.22	2.11	2.07
America	29.32	26.23	23.26	20.68	19.26
Brazil	15.10	14.5	12.62	11.50	10.61
USA	13.38	11.18[b]	9.60	8.11	7.60
Asia	6.64	6.65	6.23	6.05	5.90
Oceania	0.27	0.16	0.16	0.18	0.15
Africa	0.59	0.59	0.58	0.55	0.54
South Africa	0.41	0.40	0.40	0.40	0.40
World total	**40.73**	**38.30**	**34.71**	**31.89**	**29.81**

Notes:
[a] Figures rounded up.
[b] Includes installed capacity.
Source: F. O. Lichts (2004) p134.

diesel in the short to medium term, either blended or neat. Ethanol is particularly promising, with an estimated world production in 2004 of about 41 Bl (billion litres) (see Table 1.7). This could easily reach 60 Bl before the end of the decade, given the increasing worldwide interest, with over 30 countries in the process of implementing or planning the use of ethanol for fuel. Other estimates indicate that by 2020 ethanol could provide between 3 and 6 per cent of gasoline, approximately 129 Bl (80 Bl gasoline equivalent) or even as much as 10 per cent.

The development of bi-fuel (flex-fuel) or even tri-fuel engines is part of a new fuel trend in the transport sector. This technology does not represent any revolutionary or fundamental change, but a long chain of small improvements with potentially major impacts. In a very short time this technology has been revolutionized, with improvements continuously being added. It is particularly significant that these innovations have been implemented at low cost. There is no doubt that this technology will improve significantly in the near future and that current engine and fuel difficulties will be solved satisfactorily. The flexibility of the flex-fuel is evident, with its sophisticated system allowing the simultaneous utilization of different fuels blended in any proportion, and represents new opportunities and challenges for the industry, the consumers and society as a whole (Rosillo-Calle and Walter 2006).

Linkages between traditional and modern applications

It is difficult to predict just how long the shift from traditional to modern and efficient applications of bioenergy will take, or the exact technologies that will be used, given the many variable and complex factors involved, many of which are not directly related to energy. It is clear, however, that there is a long way to go and that this shift will be uneven (geographically, technically, socially, etc.) due to the differences in the degree of development and the strength of concern about environmental sustainability. This is particularly the case among developing nations, which differ so widely in the level of socio-economic and technological development, not to mention natural resource endowment. Furthermore, the many forms of bioenergy (solid, liquid, gaseous), the diversity of sectors involved and the many different applications will ensure that the transfer to modern bioenergy technologies will be uneven and complex.

Throughout the world many households and cottage industries already use both biomass and fossil fuels, depending on availability and price. Reliability, social status and convenience also play an important part in the choice of energy. What is important, however, is that bioenergy can increasingly be associated with modernization and environmental sustainability.

An encouraging trend is the growth in international trade in bioenergy, which until very recently took place mostly at the local level. Biotrade could bring many benefits, since it will increase competition and bring new opportunities to rural communities that have substantial natural resources and are close enough to good transportation networks. It is difficult to provide any reliable figures, but preliminary data from FAOSTAT show that in 2001–2002 there was international trading of some 1.35 Mt of charcoal, 26.74 Mm^3 of wood chips and particles, 1.93 Mm^3 of fuelwood, and 6.30 Mm^3 of wood residues.[5] (See the case study on Biotrade in Chapter 7.)

BARRIERS

The large-scale utilization of bioenergy still faces many barriers, ranging from socio-economic, cultural and institutional to technical. These barriers have been extensively investigated in the literature (e.g. Bhattacharya, 2004; G8 Renewable Energy Task Force (Anon, 2001), among others). Sims (2002) in his book *The Brilliance of Bioenergy*, identified a number of barriers and public concerns that need to be overcome, including:

- the possible destruction of native forests due to increasing commercial applications of bioenergy
- the perceived problems with dioxins
- the possible deleterious effects of large-scale dedicated energy plantations on water resources
- possible effects on soils by continuous removal of residues for energy
- potential monoculture problems posed by large-scale energy plantations (e.g. effects on biodiversity)
- possible effect of transporting large quantities of biomass (increased traffic)

- the perceived view of competition for land between food and fuel crops
- the problem of maintaining high productivity sustainably for very long periods.

These barriers should be taken into account when assessing the implications.

Technology options

Many studies have demonstrated that just minor technology improvements could increase the efficiency of biomass energy production and use significantly, maintain high productivity of biomass plantations on a sustainable basis and mitigate environmental and health problems associated with biomass production and use. The main technology options are summarized in Table 1.8, and are briefly described below and in the section on 'Secondary fuels (liquid and gaseous)' in Chapter 4.

Combustion

Combustion technologies produce about 90 per cent of the energy from biomass, converting biomass fuels into several forms of useful energy, e.g. hot air, hot water, steam and electricity. Commercial and industrial combustion plants can burn many types of biomass ranging from woody biomass to MSW. The simplest combustion technology is a furnace that burns the biomass in a combustion chamber. Biomass combustion facilities that generate electricity from steam-driven turbine generators have a conversion efficiency of between 17 and 25 per cent. Cogeneration can increase this efficiency to almost 85 per cent. Large-scale combustion systems use mostly low-quality fuels, while high-quality fuels are more frequently used in small application systems.

The selection and design of any biomass combustion system is primarily determined by the characteristics of the fuel, environmental constraints, cost of equipment and the size of plant. The reduction of emissions and efficiency are major goals (see www.ieabioenergy-task32.com/handbook.html).

There is increasing interest in wood-burning appliances for heating and cooking. Domestic wood-burning appliances include fireplaces, heat-storing stoves, pellet stoves and burners, central heating furnaces and boilers, etc.

There are various industrial combustion systems available, which, broadly speaking, can be defined as fixed-bed combustion (FxBC), fluidized bed combustion (FBC) and dust combustion (DC).

Table 1.8 *Summary of the main characteristics of technologies under consideration*

Conversion technology	Biomass type	Example of fuel used	Main product	End-use	Technology status	Remarks
1 Combustion	Dry biomass	Wood logs, chips and pellets, other solid biomass, chicken litter	Heat	Heat and electricity (steam turbine)	Commercial	Efficiencies vary e.g. >15–40% electrical; >80% thermal
2 Co-firing	Dry biomass (woody and herbaceous)	Agro-forestry residues (straw, waste)	Heat/electricity	Electricity and heat (steam turbines)	Commercial (direct combustion). Demonstration stage (advanced gasification and pyrolysis)	Large potential for use of various types of biomass; reduced pollution, lower investment costs. Some technical, supply and quality problems
3 Gasification	Dry biomass	Wood chips, pellets and solid waste	Syngas	Heat (boiler), electricity (engine, gas turbine, fuel cell, combined cycles), transport fuels (methanol, hydrogen)	Demonstration to early commercial stage	Advanced gasification technologies offer very good opportunities for using a range of biomass sources for different end-uses
4 Pyrolysis	Dry biomass	Wood chips, pellets and solid waste	Pyrolysis oil and by-products	Heat (boiler), electricity (engine)	Demonstration to early commercial stage	Issues remain with quality of pyrolysis oil and suitable end-uses
5 CHP	Dry biomass, biogas	Straw, forest residues, wastes, biogas	Heat and electricity	Combined use of heat and electric power (combustion and gasification processes)	Commercial (medium to large scale) Commercial demonstration (small scale)	Political priority in the UK, high efficiency, e.g. c.90%; potential for fuel cell applications (small plants)
6 Etherification/ pressing	Oleaginous crops	Oilseed rape	Biodiesel	Heat (boiler), electricity (engine), transport fuel	Commercial	High costs
7 Fermentation/ hydrolysis	Sugar and starches, cellulosic material	Sugarcane, corn, woody biomass	Ethanol	Liquid fuels (e.g. transport) and chemical feedstock	Commercial. Under development for cellulosic biomass	Cellulosic 5–10 year for commercialization
8 Anaerobic digestion	Wet biomass	Manure, sewage sludge, vegetable waste	Biogas and by-products	Heat (boiler), electricity (engine, gas turbine, fuel cells)	Commercial, except fuels cells	Localized use

Source: Rosillo-Calle (2003).

Co-firing

Co-firing is potentially a major option for the utilization of biomass, if some of the technical, social and supply problems can be overcome. Co-firing of biomass with fossil fuels, primarily coal or lignite, has received much attention particularly in Denmark, the Netherlands and the United States. For example, in the United States tests have been carried out on over 40 commercial plants and it has been demonstrated that co-firing of biomass with coal has the technical and economic potential to replace at least 8 GW of coal-based generation capacity by 2010 and as much as 26 GW by 2020, which could reduce carbon emissions by 16–24 MtC (Millions tonnes Carbon). Since large-scale power boilers range from 100 MW to 1.3 GW, the biomass potential in a single boiler ranges from 15 to 150 MW (ORNL, 1997).

Biomass can be blended with coal in differing proportions, ranging from 2 to 25 per cent or more. Extensive tests show that biomass energy could provide, on average, about 15 per cent of the total energy input with modifications only to the feed intake systems and the burner.

The main advantages of co-firing include:

- existence of an established market particularly for CHP
- relatively smaller investment compared to a biomass only plant (i.e. minor modification in existing coal-fired boiler)
- high flexibility in arranging and integrating the main components into existing plants (i.e. use of existing plant capacity and infrastructure)
- favourable environmental impacts compared to coal-only plants
- potentially lower local feedstock costs (i.e. use of agro-forestry residues and energy crops, if present, productivity can increase significantly)
- potential availability of large amounts of feedstock (biomass/waste) that can be used in co-firing applications, if supply logistics can be solved
- higher efficiency for converting biomass to electricity compared to 100 per cent wood-fired boilers (for example, biomass combustion efficiency to electricity would be close to 33–37 per cent when fired with coal)
- planning consent is not required in most cases.

Gasification

Gasification is one of the most important research, development and demonstration (RD&D) areas in biomass for power generation, as it is the main alternative to direct combustion. Gasification is an endothermal conversion technology in which a solid fuel is converted into a combustible gas. The importance of this technology lies in the fact that it can take advantage of advanced turbine designs and heat-recovery steam generators to achieve high energy efficiency.

Gasification technology is not new; the process has been used for almost two centuries. For example, in the 1850s much of London was illuminated by 'town gas', produced from the gasification of coal. There are now over 90 gasification plants and over 60 manufacturers around the world. The main attractions of gasification are:

- higher electrical efficiency (e.g. 40 per cent or more compared with combustion 26–30 per cent), while costs may be very similar
- important developments on the horizon, such as advanced gas turbines and fuel cells
- possible replacement of natural gas or diesel fuel used in industrial boilers and furnaces
- distributed power generation where power demand is low
- displacement of gasoline or diesel in an internal combustion (IC) engine.

There are many excellent reviews of gasification, for example Kaltschmitt and Bridgwater, 1997; Kaltschmitt et al, 1998; Walter et al, 2000.

Pyrolysis

The surge of interest in pyrolysis stems from the multiple products obtainable from this technology: for example, liquid fuels that can easily be stored and transported, and the large number of chemicals (e.g. adhesives, organic chemicals and flavouring) that offer good possibilities for increasing revenues.

A considerable amount of research has gone into pyrolysis in the past decade in many countries (see, for example, Kaltschmitt and Bridgwater, 1997). Any form of biomass can be used, although cellulose gives the highest yields at around 85–90 per cent wt on dry feed. Liquid oils obtained from pyrolysis have been tested for short periods on gas turbines and engines with some initial success, but long-term data are still lacking.

Combined heat and power (CHP)

CHP is a well-understood technology that is over a century old. Many manufacturing plants operated CHP systems in the late 19th century, although most abandoned the technology when utility monopolies began to emerge. Essentially, CHP is usually implemented by the addition of a heat exchanger that absorbs the exhaust heat, which is otherwise wasted, from an existing generator. The energy captured is then used to drive an electrical generator.

CHP is becoming fashionable primarily for the following reasons:

- *Energy efficiency* – CHP is about 85 per cent efficient, compared to the 35–55 per cent of most traditional electricity utilities.

- *Growing environmental concerns* – it is estimated that each MWe of CHP saves approximately 1000 t/C/yr.
- *Energy decentralization* – recent world market projections indicate that the market for generators below 10 MW could represent a significant proportion of the 200 GW of new capacity expected to be added by 2005 worldwide.

There is a large body of literature on CHP, for example see www.eeere. energy.gov, of the USDOE.

The future for bioenergy

Global changes in the energy market, particularly decentralization and privatization, have created new opportunities and challenges for both RE in general and bioenergy in particular. Experiments in market-based support are changing the way we look at energy production and utilization.

It is notoriously difficult to forecast long-term energy demand. However, it seems clear that demand will continue to grow. Thus, the question is, how can this demand be met and what will be the most important resources? More specifically, what is likely to be the role of bioenergy? Is RE, and more specifically biomass energy, finally reaching maturity?

Globally there is a growing confidence that RE in general is maturing rapidly in many areas of the world and not just in niche markets. It is important to recognize that the development of biomass energy will largely be dependent on the development of the RE industry as a whole, as it is driven by similar energy, environmental, political, social and technological considerations.

The 1970s were pioneering years providing a wealth of innovative ideas on RE which were further advanced in the 1980s when the computer revolution played a key enabling role. In the 1990s, improvements in RE allowed the technology to meet emerging market opportunities, such as gasification, cogeneration/CHP, etc. This opportunity was very much linked to the growing concern about climate change and the environment. The early part of the 21st century may be dominated by a global policy drive to mitigate climate change. It is essential that biomass energy is integrated with existing energy sources and thereby able to meet the challenges of integration with other RE and fossil fuels.

For bioenergy to have a long-term future, it must be produced and used sustainably to demonstrate its environmental and social benefits in comparison to fossil fuels. The development of modern biomass energy systems is still at a relatively early stage, with most of the R&D focusing on the development of fuel supply and conversion routes that minimize environmental impacts. Although the technologies are evolving quite rapidly, the R&D devoted to bioenergy is insignificant compared to that on fossil fuels, and needs to be substantially increased. In addition, the development of biomass energy should be more closely integrated with other RE technologies, and with local capacity building, financing and the like.

Modernizing bioenergy will bring many benefits. Lugar and Woolsey (1999) wrote:

> *Let us imagine, for example, that cellulose-based ethanol becomes a commercial reality. Imagine if hundred of billions of dollars that currently flows into the coffers of a handful of nations, were to flow into millions of farmers, most countries would see substantial economic and environmental benefits as well as increased national security. With so many millions involved in production of ethanol fuel, it would be impossible to create a cartel. With new drilling oil technology, we will be able to make better use of existing resources and accelerate production, but would not be able to expand oil reserves.*

The transportation system is more complex. The IC engines and oil-derived fuels have dominated the transportation systems for many decades, and have been so successful that until recently prospects for radical alternatives were not taken seriously and thus little RD&D had been directed into the search for new alternatives.

It is only in recent years that a combination of technological, environmental and socio-economic changes is forcing the search for new alternatives that could challenge the dominance of the IC engine. However, it is still unclear which alternative(s) will prevail, given the present stage of development and the range of alternatives under consideration from which no clear winners have emerged so far. In the short term, the main challenge will be to find sound alternatives to fossil fuels that can be used in the IC engine, such as ethanol or biodiesel, currently in commercial use, while others such as hydrogen are emerging. In the longer term the challenge will be to find large-scale alternatives to fossil fuels that can be used in both existing IC engines and new propulsion systems.

Data on biomass production and use remain poor, despite considerable efforts to improve the situation. Consumption data often deal with the household sector, for example excluding data on many small enterprises. In particular, the modernization of biomass energy use requires a good information base. Only a handful of countries have a reasonably good database on biomass supply, although mostly based on commercial forestry data rather than bioenergy.

Despite increased recognition, biomass energy does not receive the attention it deserves from policy makers and even less from educators. The following quotation is a good illustration:

> *Wood energy, like the oldest profession, has been around since time immemorial, like prostitution, it is ignored or regarded as an embarrassment by many decision makers at the national and international*

level. However, for about half of the world's population it is a reality and will remain so for many decades to come. (Openshaw, 2000)

NOTES

1   See for example *Renewable Energy World*, vol 7, no 4, pp238 ff.
2   This could change if carbon trading (CT) takes off on a large scale.
3   Another reason why bioenergy continues to grow, despite the fact that poor people (women and children) spend a considerable amount of time gathering firewood, dung, etc., is because biomass energy resources continue to be free in most cases. 'Free' means that poor people do not pay cash for the resource; the time spent collecting wood, etc. is not taken into account or given economic value.
4   In June 2004, representatives from 154 countries met in Bonn, Germany, and recognized the importance of RE sources and the need to promote them. It was the first time that RE received such worldwide recognition.
5   Data provided by B. Hillring, Department of Bioenergy, Swedish University of Agricultural Sciences, Uppsala, Sweden.

REFERENCES AND FURTHER READING

Anon (2001) 'G8 renewable energy task force', Final Report, IEA, Paris

Anon (2003) 'World ethanol production powering ahead', F. O. Lichts, vol 1, no 19, p139, www.agra.-net.com

Anon (2004) 'World ethanol and biofuels report', F. O. Lichts, vol 7, no 3, pp129–135

Azar, C., Lindgren, K. and Anderson, B. A. (2003) 'Global energy scenarios meeting stringent CO_2 constraints: Cost-effect fuel choices in the transportation sector', *Energy Policy*, vol 31, pp961–976

Bauen, A., Woods, J. and Hailes, R. (2004) 'Bioelectricity vision: Achieving 15% of electricity from biomass in OECD countries by 2020', WWF, Brussels, Belgium, www.panda.org/downloads/europe/biomassreportfinal.pdf

Bhattacharya, S. C. (2004) 'Fuel for thought: The status of biomass energy in developing countries', *Renewable Energy World*, vol 7, no 6, pp122–130

Bhattacharya, S. C., Attalage, R. A., Augusto, L. M., Amur, G. Q., Salam, P. A. and Thanawat, C. (1999) 'Potential of biomass fuel conservation in selected Asian countries', *Energy Conversion and Management*, vol 40, pp1141–1162

Faaij, A. P. C., Schlamadinger, B., Solantausta Y. and Wagener M. (2002) 'Large Scale International Bio-Energy Trade', Proceed. 12th European Conf. and Technology Exhibition on Biomass for Energy, Industry and Climate Change Protection, Amsterdam, 17–21 June

FAO/WE (2003) *Bioenergy Terminology*, FAO Forestry Department, FAO, Rome

Hall, D. O. and Rao, K. K. (1999) *Photosynthesis*, 6th edn, Studies in Biology, Cambridge University Press

Hall, D. O. and Overend, R. P. (1987) 'Biomass forever' in *Biomass: Regenerable Energy*, Hall, D. O. and Overend, R. P. (eds), John Wiley & Sons, pp469–473

Hall, D. O., House, J. I. and Scrase, I. (2000) 'Overview of biomass energy', in *Industrial Uses of Biomass Energy – The Example of Brazil*, Rosillo-Calle, F., Bajay, S. and Rothman, H. (eds), Taylor & Francis, London, pp1–26

Hall, D., Rosillo-Calle, F. and Woods, J. (1994) 'Biomass utilisation in households and industry: Energy use and development', *Chemosphere*, vol 29, no 5, pp1099–1119

Hemstock, S. and Hall, D. O. (1993) 'Biomass energy flows in Zimbabwe' (submitted for publication)

Hoogwijk, M., den Broek, R., Bendes, G. and Faaij, A. (2001) 'A review of assessments on the future of global contribution of biomass energy', in *1st World Conf. on Biomass Energy and Industry*, Sevilla, James & James, London, Vol II, pp296–299

IEA (2002) *Energy Outlook 2000–2030*, IEA, Paris (www.iea.org)

IEA (2002) *Handbook of Biomass Combustion and Co-firing*, Internet, International Energy Agency (IEA), Task 32 (www.ieabioenergy-task32.com)

IPCC-TAR (2001) *Climate Change 2001: Mitigation*, Davidson, O. and Metz, B. (eds), Third Assessment of the IPCC, Cambridge University Press

Kaltschmitt, M. and Bridgwater, A. V. (eds) (1997) *Biomass Gasification and Pyrolysis: State of the Art and Future Prospects*, CPL Press, Newbury, 550pp

Kaltschmitt, M., Rosch, C. and Dinkelbach, L. (eds) (1998) *Biomass Gasification in Europe*, EC Science Research & Development, EUR 18224 EN, Brussels

Kartha, S., Leach, G. and Rjan, S. C. (2005) *Advancing Bioenergy for Sustainable Development: Guidelines for Policymakers and Investors*, Energy Sector Management Assistance Programme (ESMAP) Report 300/05, The World Bank, Washington, DC

Licht, F. O. (2004). *World Ethanol and Biofuels Report*, vol 3, no 17, pp130–135

Lugar, R. G. and Woolsey, J. (1999) 'The New Petroleum', *Foreign Affairs*, vol 78, no 1, pp88–102

Night, B. and Westwood, A. (2005) 'Global growth: The world biomass market', *Renewable Energy News*, vol 8, no 1, pp118–127

Openshaw, K. (2000) 'Wood energy education: An eclectic viewpoint', *Wood Energy News*, vol 16, no 1, pp18–20

ORNL (1997) 'Potential impacts of energy-efficient and low-carbon technologies by 2010 and beyond', Report No. LBNL-40533 or ORNL/CON-444, Oak Ridge National Laboratory, Oak Ridge, TN, USA

Perlin, J. and Jordan, P. (1983) 'Running out: 4200 years of wood shortages', *The Convolution Quarterly* (Spring), Sausalito, CA 94966

Rosillo-Calle, F. (2003) 'Public dimension of renewable energy promotion: Sitting controversy in biomass-to-energy development in the UK', EPSRC Internal Report, Kings College, London

Rosillo-Calle, F. (2006) 'Biomass energy', in Landolf–Bornstein Handbook, vol 3, *Renewable Energy*, Chapter 5 (forthcoming)

Rosillo-Calle, F. and Hall, D. O. (2002) 'Biomass energy, forests and global warming', *Energy Policy*, vol 20, pp124–136

Rosillo-Calle, F. and Moreira, J. R. (2006) *Domestic Energy Resources in Brazil: A Country Profile on Sustainable Energy Development*, International Atomic Energy (IAEA/UN), Vienna (in press)

Rosillo-Calle, F. and Walter, A. S. (2006) 'Global market for bioethanol: Historical trends and future prospects, *Energy for Sustainable Development*, vol 10, no 1, pp20–32 (March special issue)

Ross, M. H. and Cobb, T. B. (1985) 'Biomass Flows in the United States Economy', Argonne National Laboratory, Argonne, IL 60439; Report ANL/CNSV-TM-172

Schubert, H. R. (1957) *History of the British Iron and Steel Industry*, Routledge & Kegan Paul, London

Senelwa, K. A. and Hall, D. O. (1993) 'A biomass energy flow chart for Kenya', *Biomass and Bioenergy*, vol 4, pp35–48

Sims, R. E. H. (2002) *The Brilliance of Bioenergy: In Business and in Practice*, James and James, London

Walter, A. S. et al (2000) 'New technologies for modern biomass energy carriers', in *Industrial Uses of Biomass Energy: The Example of Brazil*, Rosillo-Calle, F., Bajay, S. and Rothman, H. (eds), Taylor & Francis, London, pp200–253

Woods, J. and Hall, D. O. (1994). *Bioenergy for Development: Technical and Environmental Dimensions*, FAO Environment and Energy Paper 13. FAO, Rome

Zervos, A., Lins, C. and Schafer, O. (2004) 'Tomorrow's world: 50% renewables scenarios for 2040', *Renewable Energy World*, vol 7, no 4, pp238–245, www.erec-renewables.org

Appendix 1.1

Table 1.9 *Role of biomass energy by major region in 2000 (EJ/yr)*

	World	OECD	Non-OECD	Africa	Latin America	Asia
Primary energy[a]	423.3	222.6	200.7	20.7	18.7	93.7
of which biomass (%)	10.8	3.4	19.1	49.5	17.6	25.1
Final energy[a]	289.1	151.2	137.9	15.4	14.6	66.7
of which biomass (%)	13.8	2.5	26.3	59.6	20.3	34.6
Estimated modern bioenergy[b]	9.8	5.2	4.6	1.0	1.9	1.5
as % of primary energy	2.3	2.3	2.3	4.7	10.0	1.6
Modern biomass inputs to:						
electricity, CHP, heat plants	4.12	3.72	0.39	0	0.14	0.07
as % of total sector inputs	2.7	4.1	0.6	0	3.4	0.2
Industry (approx.)	5.31	1.34	3.97	0.98	1.45	1.44
as % of total sector inputs	5.8	3.0	8.6	30.3	26.0	6.3
Transport	0.35	0.10	0.26	0	0.29	0.03
as % of total sector inputs	0.5	0.2	1.1	0	6.3	0.4

Notes:
[a] Primary energy (see source for further details).
[b] Estimated modern bioenergy is the sum of bioenergy inputs to the three sectors shown in the table: electricity, CHP and district heating plant; industry (which might include some traditional bioenergy); and transport fuels.
Source: Kartha et al (2005).

Appendix 1.2

Table 1.10 *Energy scenarios – current and estimated energy consumption, 2001–2040*

	2001	2010	2020	2030	2040
World energy consumption (Mtoe)	10,038	11,752	13,553	15,547	17,690
Biomass	1080	1291	1653	2221	2843
Hydro (large)	223	255	281	296	308
Hydro (small)	9.5	16	34	62	91
Wind	4.7	35	167	395	584
PV	0.2	1	15	110	445
Solar (thermal)	4.1	11	41	127	274
Solar (thermal power)	0.1	0.4	2	9	29
Geothermal	43	73	131	194	261
Marine (tidal, wave and ocean)	0.05	0.1	0.4	2	9
Total RE	1364	1683	2324	3416	4844
% RE	13.6	14.3	17.1	22.0	27.4

Note: EREC estimates show that with a robust policy RE could supply 50 per cent of the energy by 2040 and 25 per cent if significant support is provided. One weakness of this scenario is that EREC takes into account only technological developments in the RE sector and ignores similar developments in the conventional energy sector. As is well known, the conventional sector invests a far greater amount on R&DD than the RE sector and major breakthroughs/improvements are highly possible, for example, high quality gasoline and diesel which are almost pollution-free fuels, use of heavy oils (as in Canada), etc.
Source: Zervos, Lins and Schafer (2004).

General Introduction to the Basis of Biomass Assessment Methodology

Frank Rosillo-Calle, Peter de Groot and Sarah L. Hemstock

INTRODUCTION

This chapter deals with some of the main problems in measuring the use and supply of biomass, outlines the system for classifying biomass used in this handbook and sketches general methods for assessing biomass. It also briefly looks at remote sensing, biomass flow charts, units for measuring biomass (e.g. stocking, moisture content and heating values), weight versus volume, calculating energy values and finally considers possible future trends.

Despite the importance of bioenergy, there is surprisingly little reliable and detailed information on the consumption and supply of biomass, and no standardized system for measurements and accounting procedures. This serious lack of information is preventing policy makers and planners from formulating satisfactory sustainable energy policies.

The methods given in this handbook range from ways of obtaining overall estimates of bioenergy for countries or regions, through to detailed disaggregated local information. The emphasis is on traditional bioenergy applications, although modern uses of biomass for energy are also considered. None of these methods is perfect, as each has associated penalties in time, manpower and money, and each will provide differing types of information with varying degrees of accuracy. However, used appropriately, these methodologies will pinpoint critical problems in biomass production and supply. As bioenergy is very site-specific, you would need to use your own judgement according to the circumstances as to the most appropriate way to carry out your assessment.

PROBLEMS IN MEASURING BIOMASS

There are many problems associated with measuring biomass, but they generally fall into three groups:

1   the difficulty of physically measuring the biomass
2   the many different units that are used for these measurements
3   the multiple and sometimes sequential uses to which biomass is put.

The multiple uses of biomass

An assessment of the consumption and supply of biomass is very different from a similar assessment for a commercial fuel such as kerosene. While kerosene is used as a fuel for heating and lighting, biomass provides a range of essential and interrelated needs in developing countries. These benefits include not only energy but also food, fodder, building materials, fencing, medicines and more. Biomass is rarely, if ever, planted specifically for fuel: wood that is burned is often what is left over from some other process. Biomass energy should therefore always be looked at in the context of the other benefits that biomass provides, and never just from the point of view of a single sector.

Biomass products can be modified, for example, sugar cane bagasse is used as fuel, as animal feed after hydrolysis, in the construction industry and for making paper. Other forms of biomass are deliberately modified from one energy form into another, for example, wood to charcoal, dung to biogas and fertilizer, and sugar to ethanol. Thus, it may be important to measure processed biomass as an actual or a potential energy source.

It is also necessary to know the quantity of, say, woody raw material available in order to estimate the quantity used to produce charcoal, pellets, etc. For example, if rice husks have a use as a boiler and kiln fuel, the annual production rate at specific sites is required in order to assess its economic use and physical availability for processing into, say, briquettes.

Measuring biomass

Problems arise in three main areas: distinguishing between the potential and actual supply, measuring variability, and the multiple units of measurement in use.

The difference between potential and actual supply

Biomass is often collected over a wide area from a range of vegetation types. Documenting the supply of biomass that is theoretically available for energy is a problem in itself, particularly if detailed data is needed. Given an accurate estimate of potential supply, the actual supply will then depend upon the access to this biomass. Topography will dictate how difficult the biomass is to collect, while local laws, traditions or customs may also restrict access to certain areas for the collection of biomass.

The next section lists some general points for consideration before embarking on a biomass assessment.

The ten commandments of biomass assessment

Below are what we call 'The Ten Possible Commandments' that may help you to avoid the worst pitfalls when dealing with biomass assessment.

1 *Never confuse consumption with need.* Need may exceed consumption if biomass is in short supply or expensive. Consumption is largely dependent on the perceived costs of a biomass resource. Costs reflect both supply and accessibility. Try to estimate 'basic energy needs', defined as the minimum energy requirement for basic activities such as cooking, heating and lighting. A developmental component may be added to allow for household and small-scale cottage industrial activities.

2 *Do not separate consumption from supply.* For convenience, the supply and consumption of biomass are treated separately in this handbook. However, data on the availability and on the use of biomass energy is often collected at the same time, even though different people may be responsible for collecting the two sets of data.

3 *Make your assumptions explicit.* Most empirical methods will entail implicit or explicit assumptions, or even decisions, on the nature or aims of the project. Any such underlying motives should be clearly stated to allow for consistent accounting and the possibility of comparing results over time.

4 *Your data requirements must be driven by your problem.* Estimates of consumption are always made to provide a basis for action. Data collection is not an end in itself. Collecting comprehensive, disaggregated data will place an enormous burden on your resources. You have to decide how detailed the data should be, and when aggregated information will be sufficient.

5 *Be aware of the danger of devoting too many resources to data collection and too few to analysis.* Analysis of the data can take time, and must be carefully planned. Ask yourself if you have the necessary resources to carry out a thoughtful analysis. If not, plan accordingly and explain the reasons why.

6 *Do not ignore it because you cannot measure it.* It is not always possible to quantify the demand for biomass. It is therefore advisable to incorporate a broad base of empirical information into the regular project reviews. Talk to the local population.

7 *Do not be beguiled by averages.* Demand figures given as averages should clearly be understood as central tendencies within a distribution. If sample

sizes are small, these average figures will be of little value. Averages and statistical information are valuable parts of the data presentation, but always treat average figures with caution.

There is often a considerable variation in energy consumption patterns, not only between countries, but also between areas separated by only a few kilometres, between and within ecological zones, and over time. These problems are further compounded where data are collected according to administrative boundaries. These rarely, if ever, match ecological zones. Confusion often arises when a comparison is attempted between data collected according to political and ecological zones.

8 *There is no single, simple solution so distrust the simple, single answer.* Energy is only one of the many uses to which biomass is put and often, energy is not the main use. The estimation of the supply and consumption of bioenergy is therefore not a simple matter, and there is no single method for drawing up an accounting system.

9 *The users are the best judges of what is good for them.* After all, a major objective of a biomass assessment is to help the consumer. A good understanding of the socio-economics, cultural practices and the needs of the community is therefore essential.

10 *Be flexible and modify any of the above if it seems sensible to do so.* Remaining flexible may prove to be the most important rule of all.

GENERAL CONSIDERATIONS WHEN MAKING A BIOMASS ASSESSMENT

The methods employed to assess biomass resources will vary, depending on:

- the *purpose* for which the data are intended;
- the *detail* required;
- the *information already available* for the particular country, region or local site.

Establishing the structure of the assessment

The following are a number of steps that will help you with your assessment.

Define the purpose and objectives
Clearly define the purpose and objectives of your assessment. Why is it needed? What does it seek to achieve?

Identify the audience
What audience you are addressing? For example, policy makers, planners and project managers require information expressed in different ways.

Determine the level of detail

Decide on the level of detail required for your data. Policy makers will require aggregated information, for example, while project personnel will probably require highly detailed, disaggregated data. Where fine detail is required for the implementation of projects, exhaustive investigation is necessary to provide a clear, unambiguous report on each type of biomass resource, and to provide a detailed analysis of the availability, accessibility, convertibility, present use pattern and future trends.

Biomass classification

Decide on a system of biomass classification. This handbook divides biomass into eight major categories, which will allow you to use similar methods of assessment and measurement for each type of biomass (see the section on 'Biomass classification' below).

Identify supply and demand

Identify critical areas in biomass supply and demand. This is essential if policy makers and planners are to make the right decisions.

Quantify existing data

What quantity of data is already available? A thorough literature search involving the coordination, assembly and interpretation of information from existing sources, including national, regional and local databases, and statistics produced by both government and non-governmental organizations (NGOs) may save a great deal of time and avoid unnecessary duplication of effort. A lot of data may already be available in the form of maps and reports, for example.

Care is needed when using data from existing sources. The Food and Agriculture Organization (FAO) is the main source of data on the supply of biomass, particularly woody biomass data, published in, for example, the World's Forest Inventories. The FAO and other agencies usually derive their information from country reports, many of which have not been updated for a considerable period of time, and which may not reflect a true picture for a variety of reasons (e.g. lack of resources and bias against bioenergy, particularly traditional applications). In addition, statistics are for the most part concerned only with commercial applications (e.g. in the case of forests, with the stem volume) compiled primarily from the assessment of industrial wood. In many developing countries, information on growing stock and yield tends to be incomplete and is generally considered inaccurate.

Most published data on wood biomass do not consider trees outside forests or woodlands, and thus ignore the fact that much collection of fuelwood takes place outside the forest. They also ignore small diameter trees, shrubs and scrub in the forest, which, along with branch wood, are one of the most important sources of fuelwood. Thus, with regard to fuel supply, large areas are neglected.

The use of fuelwood is normally an integral part of the informal economy and is never entered in the official statistics.

The lack of standard methods or units for documenting the supply or consumption of biomass may make it difficult to compare or incorporate data from previous surveys. Convert existing data to standard units to allow easy comparisons between localities whenever possible.

Determine methods of measurement

Decide how to measure biomass and which units to use. Rural women know well how much fuelwood in its different forms is needed to cook a meal. However, you need to obtain more than empirical knowledge using accurate scientific principles. Unfortunately there is no standard method for measuring biomass used for fuel. In future, the need for standardized methods for measuring bio-fuels will increase as new industrial applications using biomass come on stream, as these plants will need detailed and compatible information on the type and quantity of raw material they require.

There are various methods and techniques for measuring biomass, either by volume, weight or even length (see the section on 'Future trends' below). For some species, particularly those used commercially, techniques for assessing availability and potential supply are readily available. The commercial forestry sector traditionally measures biomass, especially woody biomass, by volume. However, biomass fuels are usually irregularly shaped objects (e.g. small branches, twigs, split wood, stalks, etc.), for which volume is an awkward method of measurement. Thus, for biomass energy the most appropriate measuring method is weight rather than volume.

For non-commercial species, and locations where a multitude of differing trees, shrubs, etc. are present, it is likely that there will be no assessment methodologies in existence. You may be able to adapt some methods used in assessments carried out in other locations, or even from the commercial forestry sector. However, you need to keep in mind that the supply and end use may be very different in other locations, and that the methods and techniques used in the commercial sector may be unsuitable.

Supply and demand analysis

Consider carrying out an analysis of both demand and supply. Where data on biomass supply are initially too difficult to obtain, the use of demand analysis data may be helpful to fill the information gaps.

Potential and actual supply

Differentiate between the potential and actual supply of biomass. Obtaining an accurate estimate of the potential supply can be a problem in itself since it depends on many factors such as topography, local laws, traditions, etc.

Time-series data
Aim to collect time-series data. Only data collected over a number of seasons, say five years, will show trends in use, and allow for climatic variation (both annual and seasonal).

Monitor results
Monitor the results of any programme. This will provide essential feedback to confirm that the biomass programme is meeting energy demands, and that it is sustainable.

Heating value
The proper measurement of the heating value of a biomass fuel is required if a reliable energy value of that biomass is to be obtained (see the section on 'Future trends' below). The energy content of biomass varies according to its moisture and ash content, which must be taken into account.

Accurate data collection
Sound, accurate data are essential in order to make sound decisions. Most published data on biomass energy usually consider only recorded fuelwood removal from forests, ignoring large areas of actual fuel supply of different types collected outside forests, such as twigs and small branches, shrubs, etc. Traditional applications are often an integral part of the informal economy and hardly enter official statistics.

The variability of biomass and the skills required to make the most efficient use of available biomass adds a further complication to estimating effective end use. However, this knowledge is only vaguely quantified, and rarely recorded. As a result, the figures for biomass consumption are rarely accurate.

A long-term goal in data collection is to produce a complete biomass flow diagram for the country or region in question (see the section on 'Estimating biomass flows' below).

Field surveys
Field surveys are frequently used to generate much needed data, but they are not without pitfalls. On-the-ground surveys are complex, time consuming and expensive. Therefore, you should consider a field survey only if other approaches prove insufficient. You should also consider how to supplement structured field surveys with, say, informal talks with the local population, as this is often the best way to collect data and to gain the confidence of the local population, and develop a degree of understanding of local cultural, social and economic practices.

It is important that you take care to design your survey to ask the right questions and to employ competent people for its implementation. Focus initially on those issues you consider to be absolutely necessary. In most cases, surveys should be carried out by a small multidisciplinary team, with the right skills and good structured questionnaires to ensure that all relevant information is entered in the questionnaire. This should not be necessary in the case of a small sample questionnaire. Field surveys are dealt with in more detail in the section 'Deciding about surveys' below.

Processed biomass

It is important to measure the supply of processed biomass such as sawmill waste, charcoal, etc., as this is an important source of energy in many areas. This is in addition to measuring the potential supply of biomass such as growing stock, annual yield of woody and non-woody biomass, annual crop production, animal residues, etc. These estimates of processed biomass should probably not pose any serious problem since you can obtain a lot of information from similar commercial activities elsewhere. For example, dedicated energy plantations in Brazil use basically the same biomass measurement techniques that are applied to other commercial plantations (see the section on 'Energy value of biomass' below).

Non-woody biomass

Non-woody biomass, particularly agricultural residues, animal waste and herbaceous crops, is a major source of energy. However, the use of non-woody biomass can be quite localized. For example, the large-scale use of dung is largely confined to a few countries such as India. The methodology used for estimating non-woody biomass depends on the type of material and the quantity of statistical data available or deducible. Accurate data, either on national, regional or local levels are required. For agricultural residues, for example, you should only be concerned with gathering data about what is being used as a fuel, not with the total non-woody biomass production on a given site. The same applies to animal wastes. Measuring the supply and consumption of dung will not be of any value unless dung is an important source of energy. However, the direct use of animal dung should not be encouraged for two main reasons: first, it may have greater value as fertilizer, and second, because it is of low energy value and poses serious health and environmental problems. Furthermore, because dung is held in low esteem, people switch to other fuels for social reasons if alternatives become available, even if they are more expensive (see Appendix 2.1).

Secondary fuels

Secondary fuels obtained from raw biomass (producer gas, ethanol, methanol, briquettes, etc.) are increasingly being used in modern industrial applications. New methods are being developed to deal with the measurement of industrial uses of bioenergy (see Chapter 4).

Fuel consumption patterns

Changing fuel consumption patterns are important indicators of changes in fuel consumption, which also point to socio-economic and cultural changes. Thus, for an accurate estimate of biomass consumption, it is important to capture these variations in fuel consumption patterns. Asking the right questions is vital to obtaining accurate data.

Modelling

A model may be defined as 'a simplified or idealized representation of a system, situation or process, often in mathematical terms devised to facilitate calculations and predictions'.

Models serve as a learning aid and a tool to analyse possible interactions in the system under study. Models do not provide accurate forecasts, particularly in extensive and diverse rural energy situations. A model's predictions depend upon the interests, experience and work view of those who design and implement it. Models can and have been used to define reality, to shape political debates and to legitimate political systems, while assuming the guise of independent decision support.

Therefore it is important to be aware that there are many types of model and that the application of modelling in biomass energy requires different approaches according to the needs of your assessment. The choice of modelling technique depends upon the data available and the policy area that is being studied.

Thus, the limited applications of modelling to rural energy have resulted in models failing to play as prominent a role as was envisaged a decade or two ago. This is because models have proved impractical in many cases, given the complex nature of biomass energy requirements, and in particular when it comes to traditional applications in rural areas. To be useful, models should be simple and inexpensive, easy to use and at the same time capable of facilitating and predicting. However, models are often designed to include too many variables, which make their application impractical. Modelling is therefore not described in any detail in this handbook. The inclusion of this brief summary on modelling is mostly for illustrative purposes.

The use of modelling in rural energy has been limited because of:

- the decentralized and diverse nature of the problem;
- difficulties in obtaining sufficient and reliable data;
- difficulties in fitting generalized frameworks for analysis of the energy use patterns;
- a lack of economic incentive for development of non-monetary resources in rural areas;
- the lack of political will to overcome these obstacles.

However, if you decide to use a model, ensure that it is clearly defined so that a sufficient number of key variables are selected for observation or measurement to avoid collection of unnecessary data. It is better to define the model in mathematical terms for woody biomass (e.g. stand growth model or yield table), as this will make it easier to identify sampling requirements for data collection, and ensure that they are estimated with adequate precision.

The choice of modelling technique depends on the data available and the policy area that is being studied. There are different types of model, which can be classified according several criteria:

- application (e.g. at planning policy or project level);
- scale (e.g. at village, regional, national level);
- objectives (e.g. demand forecasting, resource assessment, least cost supply planning, investment appraisal, economic development, environmental assessment, integrated planning);
- style (e.g. dedicated or generic, flexibility, level of integration, specific nature, use of scenarios);
- technique (e.g. which variables and interactions are endogenous, optimization methods, dynamic or static, demand or supply led, financial tools and environmental impact assessment (Smith, 1991)).

Databases

Biomass energy comprises many components in its production, conversion and use, and each of these components is further subdivided into many subcomponents. The availability of a good databank can facilitate information on many aspects of biomass. The aim of such a database should be to incorporate as much information as possible in an easily understandable and meaningful manner to allow the policy maker and energy planner to make the right decisions. Fortunately, good new databases have been set up in recent years, which, despite having a long way to go, represent a major improvement. The Internet and the increasing number of bioenergy-related networks have also facilitated the task of compiling data.

Remote sensing

Remote sensing techniques have been used for surveying the earth's surface from aircraft or satellites, using instruments to record different parts of the electromagnetic spectrum. Remote sensing has been used successfully to measure total biomass productivity.

Remote sensing involves an analysis of detailed land use patterns based on aerial photography or high-resolution satellite imagery (in particular SPOT and Thematic Mapper). Satellite photography can be used to determine the areas of dense woody biomass, but it cannot be used to give information on growing stock or annual increments. Aerial photography is more practical for fairly dense woodlands as it gives a higher definition than satellite imagery. Aerial photography may also allow the measurement of height, crown and even DBH of scattered trees. If woodlands are relatively undisturbed, and are representative of all or most age and diameter classes, then approximate volumes and weights can be obtained from crown cover. Of course, such data must be verified in field surveys. For farm trees there is no substitute for field surveys, as these trees are intensively managed. Aerial photographs could be used for estimating percentage cover and basic tree measurements, but only as an addition to ground surveys. To emphasize the point, remote sensing can form the basis for a comprehensive land use analysis. However, remote sensing data must be verified by detailed field-work, which can be costly. Remote sensing techniques have a lot to offer and are described in some detail in Chapter 6.

Land use assessment

Land use assessment is important since it can be a key factor in determining the actual biomass accessibility – one of the hardest and most essential tasks of any study of biomass (Nachtergaele, 2006).

The primary objective of a land use assessment or a land evaluation is to improve the sustainable management of land resources for the benefit of the people. Land assessment can be defined as:

> *The process of assessment of land performance when used for specified purposes involving the execution and interpretation of surveys and studies on all aspects of land in order to identify and make a comparison of promising kinds of land use in terms applicable to the objectives of the evaluation.*

Land evaluation is primarily the analysis of data about the land – its soils, climate, vegetation, etc. – and focuses on the land itself – its properties, functions and potential. It may be used for many purposes, ranging from land use planning to exploring the potential for specific land uses or the need for improved land management or for the control of land degradation.

Most current rural development is directed at alleviating economic and social problems, in particular hunger and poverty. Land evaluation is a useful tool as there is a clear focus on the people, the farmers, the rural communities and other stakeholders in the use of land resources.

There is now a growing need for land evaluation, particularly wherever the problems of farmers are caused or compounded by problems of the land, for example, soil fertility decline, erosion and increased frequency of droughts due to climatic change. However, an objective and systematic assessment of the suitability of land resources for diverse uses becomes all the more necessary as growing population pressures create conflicting demands for uses of land other than for agriculture, such as urbanization, transport, recreation and nature reserves.

Land suitability evaluation, the methodology set out in the Framework for Land Evaluation and later expanded in the Agro-ecological Zones methodology, was conceived and applied primarily in terms of sustainable biological production: crops, pastures and forestry. However, following the broader definitions of land and land resources, there is a growing need to address issues related to the capacity of the land to perform multiple economic, social and environmental functions.

Land performs a number of key, generally interdependent, environmental, economic, social and cultural functions, essential for life. The land can only provide these services if it is used and managed sustainably. In addition, when land is used for one purpose, its ability to perform other functions may be reduced or modified, leading to competition between the different functions. The land also provides services that are useful to humans and other species (e.g. water supply, carbon sequestration).

You may need to address the still widely held view that biomass energy competes directly with food production. In most cases, food and fuel are complementary to each other; but if there is a conflict with food production, you should identify it (see Figure 2.1).

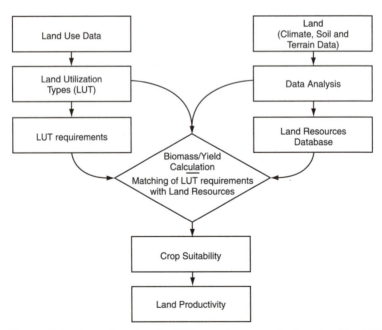

Figure 2.1 *Agro-climatic suitablility and agronomical attainable yields (Nachtergaele, 2006)*

The steps involved in a land use assessment

Assessment of land suitability is carried out by a combination of matching constraints with crop requirements, and by modelling of potential biomass production and yield under constraint-free conditions. This activity is normally carried out in two main stages, in which, first, the agro-climatic suitability is assessed and, second, the suitability classes are adjusted according to edaphic or soil constraints. Each stage comprises a number of steps, which are listed as follows:

- Stage 1: Agro-climatic suitability and agronomical attainable yields.
    1 Matching the attributes of temperature regimes to crop requirements for photosynthesis and phenology as reflected by the crop-groups, determining which crops qualify for further consideration in the evaluation.
    2 Computation of constraint-free yields of all the qualifying crops taking into account the prevailing temperature and radiation regimes.
    3 Computation of agronomically attainable yields by estimating yield reductions due to agro-climatic constraints of moisture stress, pests and diseases, and suitability of each crop in each length of growing period zone.
- Stage 2: Assessment of agro-edaphic suitability based on soil constraints.
    4 Comparison of the soil requirements of crops with the soil conditions of the soil units described in the soil inventory, at different levels of inputs.
    5 Modifications are inferred by limitations imposed by slope, soil texture and soil phase conditions.

Apart from step 2, which involves a mechanistic model of biomass production and crop yield, all the above procedures involve the application of rules which are based on the underlying assumptions that relate land suitability classes to each other, and to estimates of potential yields under different input levels. Many of these rules were derived from expert knowledge available when the first FAO Agro-ecological Zone (AEZ. See also www.iiasa.ac.at/Research/LUC/SAEZ/index.html) study was undertaken, and they should be regarded as flexible rather than rigid. The number of suitability classes, the definition of management and input levels, and the relationships between them can be modified according to increasing availability of information and the scope and objectives of each particular AEZ investigation. Results on a worldwide basis for more than 150 crop varieties were published by FAO/IIASA in 2000 and 2002. The results are available online (*www.iiasa.ac.at/Research/LUC/SAEZ/index.html*).

Changing land use

It is extremely valuable to examine the change in land use over time. This can be done in two complementary ways:

1 by analysis of historical remotely sensed data if this is available, and
2 by detailed discussion in the field with the local population.

Changes in land use patterns can be related to the issues under examination to help understand the evolution of local biomass resources.

The method of assessing land use at the planning level depends entirely on the circumstances. The national policy-level analysis should form the basis for this assessment, but further enhancement and disaggregation will almost certainly be needed. A fully disaggregated land use assessment is not normally appropriate, but a proper analysis of agro-climatic zones will be necessary. If this information is not available, some form of remote sensing is the most viable option. The particular technique chosen will depend on local circumstances. Aerial photography and a full analysis of SPOT or Thematic Mapper imagery are usually too expensive in terms of both time and money at this stage.

Conclusions on assessments

To sum up, unless detailed studies from the area in question already exist, some detailed local field work at chosen sample sites is desirable. These surveys should be minimized so that they are not too costly or demanding of time. However, they should be detailed enough to provide a true picture.

Remember, a major factor that will determine the nature of your assessment will be the financial and human resources available to you. Almost inevitably, you will have to compromise between the kind of assessment that is desirable and the assessment that is possible given the resources at your disposal.

This being the case, it is important to use your resources as effectively as possible. For example, the time you devote to the assessment of a particular biomass resource should be determined by its relative importance to consumers. Unproductive woodland does not warrant the same degree of attention as woodland from which a large amount of firewood is collected. Keep in mind consumers' preferences and other available alternatives

DECIDING ABOUT SURVEYS

Before undertaking the survey, it is necessary to consider six basic questions, as detailed below.

What is the problem?

A biomass assessment is carried out in order to implement some kind of action. The nature of the action required will determine the nature of the assessment. It is therefore essential to define the objectives of the assessment and the audience to be addressed. The nature of the problem tends to change over time and according to the nature of the biomass. For example, the problems posed by traditional applications, such as lack of firewood for cooking, are very different from those posed by modern applications, such as cogeneration.

Who is your audience?

It is convenient to think of various types of audience, for example, policy makers, energy planners and project implementers, each of whom will have different objectives and perceptions about biomass. Energy planners would have a better understanding of energy supply and demand, although if they come from the conventional energy sector, they could be biased. Understanding your audience is therefore very important in order to assist them to plan and implement biomass energy interventions.

How detailed should the information produced be?

The level of information required can only be decided once you define the reasons for the survey and the professionals to whom the results will be addressed. The type of information, and the manner in which it is presented, ranges from a generalized but succinctly presented picture for the policy maker to detailed scientific data for the project officer. Your assessment may contain any level of precision of detail between these two extremes, or it may contain sections addressed to different audiences. Whatever the case may be, the method you adopt to collect and present your data is determined in the first instance by the information needs of the people you wish to address. Your data must be a reflection of the problem you are addressing; what changes is the way you present it so that people can understand you. However, it is vital to be honest; do not hide the reality of a situation just because your audience might not like your data.

What resources are available for the survey?

The financial and human resources available are the next important factors that will determine the nature of your assessment. You will almost certainly have to compromise between the kind of assessment that is desirable, and what is possible given the resources at your disposal. Do not try to do too much if you do not have the means. Explain the situation so your audience understands your limitations.

Use your resources as effectively as possible. The time devoted to the assessment of a particular biomass resource should reflect its relative importance to consumers. Unproductive woodland does not deserve the same degree of attention as woodland from which a large amount of, say, firewood is collected.

Is a field survey necessary?

Field surveys can produce accurate and detailed data. However, they are often complex, time consuming, expensive and demanding of skilled personnel, especially in remote areas and harsh terrains. Explore alternatives such as analysis of existing data and collaboration with national surveys. Only consider mounting a field survey if there is no alternative, or if the areas involved are not too large. Field surveys should be considered only when other approaches would not be sufficient.

What is the scope and quality of existing data?

Much data is often already available, in the form of maps and reports, for example. Make a thorough search of the available literature before you begin your survey. In recent years a lot of data have been collected by many international and government agencies. Another good source is the Internet. However, care is needed when using data from existing sources. Information on the availability of woody biomass is often patchy, especially in developing countries that do not have the resources to carry out large detailed surveys. In particular, there is a pressing need to improve the statistics on fuelwood, which constitutes the bulk of the wood used in many developing countries, in terms of traditional uses/applications. Most published data on woody biomass considers only recorded (official) fuelwood removal from forest reserves (although it is often not made clear that this is the case). Do not forget that traditional uses of biomass are part of the informal economy, that therefore official consumption figures do not reflect the true picture. The official figures also ignore large areas of actual fuel supply, including:

- unrecorded removals from forests;
- trees on roadsides and community and farm lands;
- small diameter trees;
- shrubs and scrub in the forest;
- branch wood.

What equipment is needed?

Having decided on the nature and detail of the resource survey, the next step is to make an accurate assessment of a country, region or locality's natural resources. These should include:

- land type;
- vegetation type;
- soil composition;

- water availability;
- weather patterns.

Biomass classification

There are many ways of classifying biomass, but generally it can be divided into woody biomass and non-woody biomass, including herbaceous crops. The system adopted in this handbook divides biomass types into eight categories. This is attractive because it allows similar methods of assessment and measurement for each type of biomass. You may be inclined to use a more refined classification system, but whatever method you select, make sure that it is clearly specified.

1 *Natural forests/woodlands.* These include all biomass in high standing, closed natural forests and woodlands. Forests are defined as having a canopy closure of 80 per cent or more, while woodland has a canopy closure of between 10 and 80 per cent. This category will also include forest residues.

2 *Forest plantations.* These plantations include both commercial plantations (pulp and paper, furniture) and energy plantations (trees dedicated to producing energy such as charcoal, and other energy uses). The total contribution of bioenergy in the future will be strongly linked to the potential of 'energy forestry/crops plantations' since the potential of residues is more limited. In the 1970s and 1990s energy plantations were heralded as the major source of biomass energy in the future. In recent years, however, their potential has been considered to be more limited (see Chapter 1).

3 *Agro-industrial plantations.* These are forest plantations specifically designed to produce agro-industrial raw materials, with wood collected as a by-product. Examples include tea, coffee, rubber trees, oil and coconut palms, bamboo plantations and tall grasses

4 *Trees outside forests and woodlands.* These consist of trees grown outside forest or woodland, including bush trees, urban trees, roadside trees and on-farm trees. Trees outside forests have a major role as sources of fruits, firewood, etc., and their importance should not be underestimated.

5 *Agricultural crops.* These are crops grown specifically for food, fodder, fibre or energy production. Distinctions can be made between intensive, larger-scale farming, for which production figures may show up in the national statistics, and rural family farms, cultivated pasture and natural pasture.

6 *Crop residues.* These include crop and plant residues produced in the field. Examples include cereal straw, leaves and plant stems. Fuel switching can result in major changes in how people use biomass energy resources. For example, in China a rapid switch from agricultural residues to fossil fuels is causing serious environmental problems as the residues are now being burnt due to the declining market. Be aware of these pitfalls.

7   *Processed residues.* These include residues resulting from the agro-industrial conversion or processing of crops (including tree crops), such as sawdust, sawmill off-cuts, bagasse, nutshells and grain husks. These are very important sources of biomass fuels and should be properly assessed.

8   *Animal wastes.* These comprise waste from both intensive and extensive animal husbandry. When considering the supply of biomass, it is also important to ascertain the amount that is actually accessible for fuel, not the total amount produced. You need to be aware that there are large variations that can be attributed to a lack of a common methodology, which is the consequence of variations in livestock type, location, feeding conditions, etc. Animal waste may also have a better value as fertilizer. In addition, animal wastes are more frequently used for producing biogas for environmental rather than energy purposes. Your survey needs to reflect these rapidly changing uses for animal wastes, and the reasons for these changes.

LAND USE ASSESSMENT

It is advisable to begin collecting data at the most general, aggregated level. More detailed information can then be collected if necessary. Data concerning the local land use pattern will be required for project implementation, for example, when an up-to-date analysis is usually essential. Any information, whether already existing or based on new field surveys or remote sensing, should be carefully verified in the field.

An indication of the changes in land use patterns over time can provide a useful understanding of the evolution of local biomass resources and enable predictions of likely resource availability in the future. It is therefore extremely valuable to examine the change in land use over time. This can be done in various complementary ways:

- by the analysis of historical remotely sensed data (e.g. aerial photography, if this is available);
- by the use of official agricultural data;
- through detailed discussion in the field with the local population.

THE IMPORTANCE OF THE AGRO-CLIMATIC ZONE

A proper analysis of agro-climatic zones will be necessary if a thorough assessment is to be carried out. If this information is not available, some form of remote sensing is the most viable option (see Chapter 6). The particular technique chosen will depend on local circumstances. At this stage, aerial photography and a full analysis of SPOT or Thematic Mapper imagery is usually too expensive in terms of both time and money and could only be justified in large scale projects.

Variation with agro-climatic zone and with time

Remember that the biomass yield will vary with biomass type and species, agro-climatic region, rainfall, the management techniques employed in biomass production (e.g. intensive on extensive forestry or agriculture, irrigation and the degree of mechanization, etc.). These factors need careful consideration if general estimates are made in the absence of detailed land use data.

The productivity of biomass will almost certainly vary across seasons and over years. It is important to collect time-series data wherever possible. Particular care is also needed if the biomass is not homogeneous. When estimates of vegetation cover are made in the absence of detailed land use data, it is important to recognize the effect of climate, altitude and geology on plant growth. Do not forget that water is the key factor in productivity, and thus you will need to know rainfall patterns over time.

WOODY AND NON-WOODY BIOMASS

Classification into woody and non-woody biomass is often for convenience only as there is no clear-cut division between them. The way in which these biomass types are classified should not dictate which data are collected. For example, cassava and cotton stems are wood, but as they are strictly agricultural crops it is easier to treat them as non-woody plants. Bananas and plantains are often said to grow on 'banana' trees, although they are also considered to be agricultural crops. Coffee husks are treated as residues, whereas the coffee clippings and stems are classed as wood.

In some areas of developing countries tall grasses are also used for energy (e.g. cooking and heating). More recently, various grasses (e.g. miscanthus, elephant grass – see Chapter 4) are being investigated as possible sources of energy in modern commercial applications. In this case commercial measuring techniques would apply. Generally, the non-woody biomass includes the following:

- agricultural crops;
- crop residues;
- processing residues;
- animal wastes.

Woody biomass is perhaps one of the most difficult measurements to make. However, it is usually the most important form of biomass energy to document, particularly when used in traditional applications in many developing countries.

It may therefore be necessary to spend a great deal of your resources collecting data on woody biomass. Every effort should be made to determine growing stock and, in particular, the annual growth increment of standing forests (see Chapter 3).

For the purposes of policy making, an assessment of non-woody biomass can rely on statistics for agricultural production and any information available concerning agro-industrial residues. Such data will give a quick picture of the possible supply of such non-woody biomass resources as crops, crop residues and processing residues. If the required data is not available, an estimate of annual biomass production can be obtained from agricultural land use maps, in conjunction with reported figures on crop yield per unit area. Field studies and experimentation should be kept to the bare minimum.

Where finer detail is required for the implementation of projects, exhaustive investigation is necessary to provide a clear, unambiguous report on each type of biomass resource, and to provide a detailed analysis of the availability, accessibility, collectivity, convertibility, present use pattern and future trends.

For agricultural crops, dependable information on yields and stocks, quantified accessibility, material that can be collected, calorific values, storage and/or conversion efficiencies must be accurately determined. A study of the socio-cultural behaviours of the inhabitants of the project area will help to determine use patterns and future trends.

In the case of crop residues and processing residues, residue indices must also be determined in addition to the above parameters. It is also important to critically analyse the various uses of such residues. (See Appendix 2.1 Residue calculations.)

The methodology employed for estimating non-woody biomass depends on the type of material and the quality of statistical data available or deducible, and often the end use (see Chapter 4).

Alternative uses for biomass, and the proximity of the supply, also need consideration in determining the amount of woody biomass available for fuel. Wood is rarely grown specifically for fuel, because fuelwood is cheaper (sometimes a great deal cheaper) than wood sold for other purposes. An alternative use normally has priority over use of wood as fuel. Forest energy plantations are increasing (for example, there are about 3 Mha of eucalyptus plantations in Brazil and 14,000 ha of willows in Sweden used explicitly for energy), but contribute only a small proportion compared to residues (all sources). So far energy plantations have failed to take off on a large scale.

However, when trees are converted to wood products considerable waste is generated. In the forest or on the farm the buyer (or seller) only removes logs of specific dimensions. Branches and crooked stem wood may be left when the trees are felled, which can amount to between 15 and 40 per cent or more of the above-ground volume. Sawn wood may only account for between one-third and one-half of original saw logs; that is, the waste is 50–67 per cent of the log.

All the waste materials are potentially burnable, especially if they are near to the demand. Even after wood is converted into, say, poles or sawn timber it can still be used for fuel after its useful life is over. All these facts have to be considered (see Appendix 2.1).

Proximity to demand centres or markets is also very important. The nearer it is to the consumer, the more likely it is that most if not all of the biomass will be used. On the other hand, if the wood is remote from the demand, there may be a surplus of biomass that is not readily available as fuelwood or charcoal. Thus, once a biomass supply map is drawn up, it is important to match it with population densities to get a picture of how much of this biomass it is feasible to use.

It is common that large amounts of biomass remained unused because they are far away from the main consumption areas and are therefore physically or economically inaccessible. This is why it is important to consider supply to specific areas and not only look at a countrywide picture. Fuel switching is also very important. You need to assess the availability of other fuels. In a world that is urbanizing rapidly, people are switching to more convenient fuels for economic and social and cultural factors if there are other, better alternatives, even if they are more expensive.

The method of estimating woody biomass, whether direct or indirect, is determined by a number of characteristics, such as the area over which it is found, its variability and its physical size, together with the nature of the exploitation.

Past surveys have concentrated on natural forests, plantations or woodlands, as many people have the fixed idea that this is where fuelwood and timber come from. However, many demand surveys show that trees outside the forest or woodlands are a very important source of fuel, poles and even roughly hewn or hand-sawn timber. As these non-traditional sources of wood are so neglected, and because there is much more diversity of tree management in these areas, more effort should be put into measuring these trees than 'forest' trees.

Assessment of accessibility

The potential supply and the actual availability of biomass are very rarely the same. Establishing the proportion of the total biomass that is accessible, and therefore available for fuel, is one of the hardest and most essential tasks in any study of biomass supply. The variability and complexity of the factors involved with access can make quantitative analysis difficult.

Access to potential biomass resources is limited by three main physical and social constraints:

- location constraints;
- tenure constraints (social, political and cultural);
- constraints derived from the land resource management systems.

Locational constraints

Locational constraints reflect the physical difficulties of harvesting, collecting and transporting biomass from the point of production to the place where it will be burned. The gathering and transport of biomass is influenced by the terrain and the distance over which the biomass is transported, and also by the availability of biomass in a determined area. Rivers, steep slopes, areas of marshland and so on, all act as barriers to access. Locational constraints greatly influence the cost in both behaviour and financial terms of biomass energy.

It is possible to assess localized constraints cartographically, by measuring the distance between resource and consumers and noting features of the terrain. If required, more detailed information may be obtained from detailed cartographic data (from existing maps and/or a remote sensing analysis) and a programme of fieldwork. Fieldwork will provide further information on the time necessary to collect fuel, distances travelled and the points at which time and distance begin to limit accessibility. Locational constraints are an important factor needing careful analysis, as is the efficiency of the transport system.

Tenure and land management constraints

Tenure and land management constraints stem from land ownership/land rights. Land tenure patterns are highly specific to individual countries, and political and cultural aspects are very important. It is possible to identify three broad categories of land ownership:

1   Small farms owned or rented by people in the local community, where resources are subject to private property rights.
2   Communal land, owned by local people or groups/local communities.
3   Large areas of land owned and controlled by individual landowners and institutions, either commercial farms and plantations or state land such as forest reserves, game reserves, etc.

Issues of tenure do not always affect small farms and communal land unless major political and social changes are occurring. Thus, an assessment of tenure constraints should concentrate on areas where land is held by the state and the commercial farming and plantation sector. Where plantations, state reserves and the like are frequently closed off to the local community, information on accessibility is easily obtained from the institution controlling the area. Reference to these institutions will establish the policies regarding access by the local community for fuelwood collection. However, when the land is owed or used by many small farmers, the task will be more complex since many people may have to be contacted before access can be gained.

Where a detailed analysis is required, some level of fieldwork is needed to establish, for example, the extent of illicit collection. The question as to whether private property rights on small farms constrain accessibility is often a difficult question to resolve. It may be hard to determine whether a resource is limited because the supply is inaccessible or because certain sections of the community are deprived of access. Here your social and political skills will be needed!

Estimating biomass flows

A long-term goal in data collection is to produce a complete biomass flow diagram for the country, region or locality. A flow diagram traces biomass from production to end-use. It should encompass all forms of biomass production (agriculture, forestry, grasslands, etc.), allow for losses during conversion and provide details of all its uses (food, feed, timber, fuelwood, animals). However, this can be both time consuming and expensive.

To construct a flow diagram, data has to be collected systematically, starting with aggregated data, and working towards a fine level of detailed information. As flow diagrams are drawn to scale, units must be consistent throughout. Accurate flow diagrams therefore require a painstaking analysis of biomass supply and consumption. However, once established, such a diagram is a useful method of presenting data, can give an excellent national, regional and local overview, and provide an easy means of monitoring changes in the production and use of biomass.

Example of a biomass flow chart (general assumptions)

The biomass energy flow chart shown in Figure 2.2 was produced from data centred on the three main areas of biomass production, namely agriculture, forestry and livestock. The flow chart estimates total biomass energy theoretically available, its production, present utilization levels and the potential availability of biomass residues from agriculture, forestry and livestock. The flow of biomass in all its forms was followed from its production at source and harvest through to its end-use and categorized into a product or 'end-use' group (e.g. food, fuel, residues, etc.). Figure 2.2 illustrates a biomass flow chart based on the Islands of Vanuatu (see Appendix 2.2).

The following general assumptions apply to the flow chart in Figure 2.2:

- All production refers to above-ground biomass; water surfaces were not considered as components of biomass energy production.
- The following energy values (GJ t^{-1} air dry, 20 per cent moisture content) which assume direct combustion were used:
  — 1 tonne fuelwood = 15 GJ;
  — 1 tonne stemwood = 15 GJ;

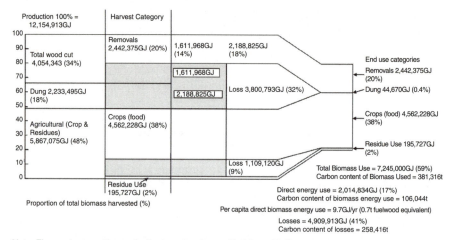

Note: Figures in parentheses indicate percentage of total production.
Source: Based on annual averages (2000–2003) FAOSTAT (www.fao.org).

Figure 2.2 *Biomass energy flow chart for Vanuatu*

— 1 tonne forest or tree harvesting residues (unmerchantable portion) = 15 GJ;

— 1 tonne charcoal = 31 GJ; efficiency of conversion for fuelwood to charcoal = 15 per cent by weight.

- All roundwood volume was assumed to be solid with a conversion equivalence of 1 m³ solid roundwood = 1.3 t.

- The term 'consumption' as used in the end-use analysis refers to the gross amount of biomass material devoted to a specific use category (e.g. food, fuelwood), while the term 'useful energy' refers to the energy content of the final state of the material. Thus 'useful energy' is 'consumption' less 'losses' during conversion.

Changing these assumptions to account for specific local/national/regional production will increase the overall accuracy of the flow chart.

What does the flow chart highlight?

Terrestrial above-ground biomass production and utilization in Vanuatu (see Appendix 2.2) was analysed for the years 2000–2003 using FAOSTAT derived data (as biomass is seasonal and responds to variable environmental factors such as rainfall you should use data over an average of at least three years). For the years 2000–2003 the total production of biomass energy was estimated at an annual average of 12 PJ (48 per cent from agricultural crop production, 34 per cent from forestry and 18 per cent from livestock). Of the 10 PJ produced from agricultural and forestry operations, 1.8 PJ of biomass was harvested and burned (106,044 t carbon equivalent), 4.5 PJ was harvested for food, 3 PJ was unutilized crop and forestry residues, 0.2 PJ was harvested crop residues for use directly as fuel. Livestock produced a further 2 PJ, of which only 0.04 PJ was harvested and used for fuel. Only 7 PJ (59 per cent) of the 12 PJ of biomass energy produced was actually utilized. A total of 3 PJ remained as unused residues and dung (from agricultural production and livestock), and a further 2 PJ was unused forestry residues. The total amount of biomass (fuelwood, residues and dung) used directly to provide energy was estimated at 2 PJ (9.7 GJ per capita per year or 0.7 t fuelwood equivalent).

Figure 2.2 highlights biomass use and areas for potential use of biomass for energy – here it is evident that losses from agriculture, forestry and livestock warrant further investigation as potential sources of biomass energy. However, it does not account for important issues such as collection efficiency, etc.

STOCK AND YIELD

The assessment of all biomass resources (woody, non-woody and animal) requires an estimation of both stock and yield. If biomass is viewed as a renewable resource, it is the annual production, or increment, that is the key factor. Stocks become depleted when the biomass harvested is greater than the increment.

Stock is defined as the total weight of biomass as dry matter.

Yield is defined as the increase in biomass over a given time and for a specific area. Yield must include all biomass removed from the area. Yields are expressed in two forms:

1 Current annual increment (CAI) – the total biomass produced over a period of one year. For annual plants, it is the total yield over the year. For perennial plants, such as trees and other woody biomass, the CAI will vary according to the season and the growing conditions. For perennials, a mean figure should be calculated from measurements made every year and in the same season.
2 Mean annual increment (MAI) – which is the total biomass produced for a certain area, divided by the number of years taken to produce it. MAI is an *average* measure of yield (see Appendix I Glossary of Terms).

Animal stock is measured as the number of animals by species.

Having determined the quantity of biomass potentially available, its energy value will depend on moisture and ash content.

Moisture content

When biomass is burned, part of the energy released is used to turn the water it contains into steam. It follows that the drier the biomass, the more energy there is available for heating.

It is therefore the moisture content that primarily determines the energy value of the biomass. Thus, while wood has a higher energy value than the other two forms of biomass at a given moisture content, it is possible for crop residues and dung to have a higher value than wood if they have lower moisture contents.

The energy value of a unit weight of biomass is inversely proportional to the amount of water it contains. To get the true weight of biomass it is therefore necessary to calculate the moisture content (mc). This can be measured in two ways: on a wet basis (wb) or a dry basis (db). These measurements are calculated as follows:

- Dry basis

$$\frac{\text{Wet weight} - \text{dry weight}}{\text{dry weight}}$$

- Wet basis

$$\frac{\text{Wet weight} - \text{dry weight}}{\text{wet weight}}$$

Air-dry wood (15 per cent mc db) has an energy value of about 16.0 MJ/kg, whereas green wood (100 per cent mc db) has a value of 8.2 MJ/kg. The energy value of oven dry woody biomass can be taken as 18.7 MJ/kg, ±5 per cent. Resinous wood has a slightly higher value and temperate hardwood a slightly lower value.

Ash content

The higher the ash content, the lower the energy value. On an 'ash free' basis, that is, when the non-combustible material is discounted, all non-woody plant biomass has more or less the same energy value.

Ash contents vary from one residue to another. For example, rice husks have 15 per cent ash content, maize cobs 1 per cent, so they have different energy values. Oven dry rice husks with a moisture content of 15 per cent are 85 per cent fibre, but as 15 per cent of this is not combustible, the husks consist of 70 per cent combustible material. Maize husks, on the other hand, at 15 per cent mc db, are 84 per cent combustible material, 17 per cent more than the rice husks.

Ash contents should be compared between biomass samples with the same moisture content.

Energy value of biomass

The energy available from biomass is expressed in two main forms:

1 gross heating value (GHV), also expressed as higher heating value (HHV), and
2 net heating value (NHV), also called low heating value (LHV).

Although for petroleum, for example, the difference between the two is rarely more than about 10 per cent, for biomass fuels with widely varying moisture contents, the difference can be very large and therefore it is very important to understand these parameters.

GHV refers to the total energy that would be released through combustion divided by the weight of the fuel. It is widely used in many countries. The NHV refers to the energy that is actually available from combustion after allowing for energy losses from free or combined water evaporation. It is used in all the major international energy statistics. The NHV is always less than GHV, mainly because it does not include two forms of heat energy released during combustion:

- the energy to vaporize water contained in the fuel, and
- the energy to form water from hydrogen contained in hydrocarbon molecules, and to vaporize it.

Calculating energy values

For example, for zero-moisture wood the energy value is 20.2 MJ/kg, for crop residues 18.8 MJ/kg and for dung 22.6 MJ/kg. The difference between high and low heat values is approximately 1.3 MJ/kg at 0 per cent mc. This figure is therefore deducted from the high heat value to obtain the low heat values of 18.9 MJ/kg for wood, 17.6 MJ/kg for crop residues and 21.3 MJ/kg for dung. These values are then used to calculate the low heat values at different moisture and ash contents. Wood with a moisture content of 80 per cent db contains 44 per cent water and 56 per cent fibre. If all the fibre is burnable, the energy content is $0.56 \times 18.9 = 10.6$ MJ. However, 1 per cent is non-combustible, so the energy content is $0.56 \times 18.9 \times 0.99 = 10.5$ MJ/kg. Some of this energy is required to drive off the water. To expel 0.44 kg of water will take $2.4 \times 0.44 = 1.1$ MJ (heat required to drive off 1 kg of water). Thus the net energy available for heating is $10.5 - 1.1$ MJ/kg. This formula can be used if the high heat value or the low heat value is known at specific moisture and ash contents. (See Appendix 2.3.)

Furthermore, the difference between NHV and GHV depends largely on the water (and hydrogen) content of the fuel. Petroleum fuels and natural gas contain little water (3–6 per cent or less) but biomass fuels may contain as much as 50–60 per cent water at point of combustion. Heating values of biomass fuels are often given as the energy content per unit weight or volume at various stages: green, air-dried and oven-dried material (see Appendix I Glossary of Terms).

Many surveys go into elaborate detail about energy values of fuel and record them to several decimal places. As data are usually only accurate to within, at best, 20 per cent, such detail is generally unrealistic. The net amount of energy available from biomass as heat depends upon two factors:

1   the amount of water it contains;
2   the quantity of non-combustible material in the biomass which will be left as ash after burning (the ash content).

The substances that form the ashes generally have no energy value. For woody biomass, the ash content is more or less constant at around 1 per cent for all species. It is therefore the moisture content rather than the species of wood that determines energy availability.

For non-woody biomass, ash content can be more important.

Dung

The ash-free energy value of animal dung is higher than that of wood. Oven dry dung has a low heat value of about 21.2 MJ/kg. At 15 per cent mc db this is reduced to 18.1 MJ/kg. However, with an average ash content ranging from 23–27 per cent, the actual energy value at 15 per cent mc db is about 13.6 MJ/kg.

Crop residues

On an ash-free basis, the energy value of crop residues is slightly less than that of wood, principally because they have a lower carbon content (about 45 per cent) and higher oxygen content.

Ash content varies from one residue to another. For example, rice husks have 15 per cent ash content, maize cobs 1 per cent – so they have different energy values. The average energy value of ash-free, oven-dry annual plant residues is about 17.6 MJ/kg. At 15 per cent mc db, the energy value of the ash-free residue is about 15.0 MJ/kg. With a 2 per cent ash content, the energy value will be about 14.7 MJ/kg, and with a 10 per cent ash content, about 13.5 MJ/kg.

Always compare the ash content of different biomass samples that have the same moisture content. Oven-dry rice husks with a moisture content of 15 per cent are 85 per cent fibre, but as 15 per cent of this is not combustible, the husks consist of 70 per cent combustible material. Maize husks on the other hand, at 15 per cent mc db, are 84 per cent combustible material, 17 per cent more than the rice husks.

Charcoal

The energy value of charcoal not only depends on the moisture and ash contents, as with other forms of biomass, but also on the degree of carbonization. Charcoal is obtained by the carbonization (pyrolysis) of wood by heat in the absence of air at a temperature above 300 °C, when volatile components of wood are eliminated. During the process, there is an accumulation of carbon in the charcoal from about 50 per cent to about 75 per cent, due partly to the reduction of H and O in the wood. The average moisture content of charcoal is about 5 per cent (db). It will absorb water only gradually unless deliberately wetted, so the moisture content can be treated as constant.

The ash content of charcoal depends on its parent material. Wood charcoal may have up to 4 per cent ash content and coffee husk charcoal 20–30 per cent. Thus, assuming full carbonization, the wood charcoal will have about 33 per cent more energy than coffee husk charcoal per unit weight at the same moisture content. Tests should be undertaken to determine both the moisture content and the ash content of charcoal (plus soil and other foreign body content if necessary).

Weight versus volume

The forest industry measures wood by volume, but biomass fuels should be measured by weight. This is because the heating value, or amount of heat that can be provided, must be referred to on a weight basis. Weight estimates can be obtained directly from measurements of tree dimensions, or indirectly via wood volume measurements. Direct determination of weight is preferable for biomass energy supply assessment.

Biomass that is traded – timber and agricultural commodities – is measured in standard units suited to a particular commodity. For example, foresters measure timber by volume, because they are concerned with the bulky and more-or-less uniform stems and trunks. However, biomass fuels are often irregularly shaped (twigs, small branches, leaves, stalks, etc.), making volume an awkward method of measurement. In addition, the weight is required when determining the heat value of biomass. Biomass for energy should always be measured by weight.

FUTURE TRENDS

You must be aware of changing trends in measuring techniques, fuel switching, social, economic and policy changes, potential energy alternatives and ways and means of enhancing existing resources in an environmentally sustainable manner. The environmental implications so often ignored in the past, must be fully taken on board.

For example, it is essential to be aware of the potential for increasing yields from existing and new species or clones. This knowledge should be based on national and international research and experience. You must also appreciate the conflicting pressure of land use trends and socio-economic changes that will affect the possibility of maintaining (or increasing) biomass supplies. Such information needs to be continuously made available at the planning and policy levels, otherwise decisions take on the ground may be very difficult to implement.

The aim is to achieve optimal and sustained productions of biomass in a manner that fulfils both environmental and socio-economic criteria, to allow the policy maker to make sound policy decisions.

REFERENCES AND FURTHER READING

Bialy, J. (1986) *A New Approach to Domestic Fuelwood Conservation: Guidelines for Research*, FAO, Rome

Hall, D. O. and Overend, R. O. (eds) (1987). *Biomass: regenerable energy*, John Wiley & Sons, Chichester, UK

Hall, D., Rosillo-Calle, F. and Woods, J. (1994) 'Biomass utilization in households and industry: Energy use and development', *Chemosphere*, vol 29, no 5, pp1099–1119

Hemstock, S. L. and Hall, D. O. (1994) 'A methodology for drafting biomass energy flow charts', *Energy for Sustainable Development*, 1, pp38–42

Hemstock, S. L. and Hall, D. O. (1995) 'Biomass energy flows in Zimbabwe', *Biomass and Bioenergy*, 8, pp151–173

Hemstock, S., Rosillo-Calle, F. and Barth, N. M. (1996) 'BEFAT – Biomass Energy Flow Analysis Tool: A multi-dimensional model for analysing the benefits of biomass energy', in *Biomass for Energy and the Environment*, Proc. 9th European Energy Conference, Chartier et al (eds), Pergamon Press, pp1949–1954

Hemstock, S. L. (2005) *Biomass Energy Potential in Tuvalu* (Alofa Tuvalu), Government of Tuvalu Report

Kartha, S., Leach, G. and Rjan, S. C. (2005) *Advancing Bioenergy for Sustainable Development; Guidelines for Policymakers and Investors*, Energy Sector Management Assistance Programme (ESMAP) Report 300/05, The World Bank, Washington, DC

Leach, G. and Gowen, M. (1987) *Household Energy Handbook: An Interim Guide and Reference Manual*, World Bank Technical Paper No 67, World Bank, Washington, DC, pp16–20

Nachtergaele, F. (2006). FAO Land and Water Development Divison, Rome (information supplied by Freddy Nachtergaele)

Ogden, J., Williams, R. H. and Fulmer, M. E. (1991). 'Cogeneration applications of biomass gasifier/gas turbine technologies in cane sugar and alcohol industries'. In *Energy and the Environment in the 21st Century*, edited by J. W. Tester, D. O. Wood and N. A. Ferrari, MIT Press, Cambridge, Massachusetts, pp311–346

Rosillo-Calle, F., Furtado, P., Rezende, M. E. A. and Hall, D. O. (1996) *The Charcoal Dilemma: Finding Sustainable Solutions for Brazilian Industry*, Intermediate Technology Publications, London

Rosillo-Calle, F. (2001) *Biomass Energy (Other than Wood) Commentary 2001*, Chapter 5: Biomass, World Energy Council, London

www.fao.org/waicent/portal/statistics_en.asp

Smith, C. (1991) 'Rural Energy Planning: Development of a Decision Support System and Application in Ghana', PhD Thesis, Imperial College of Science, Technology and Medicine, University of London.

World Resources Institute (1990) *World Resources 1990–1991: A Guide to the Global Environment.* Oxford University Press, Oxford, UK

APPENDIX 2.1 RESIDUE CALCULATIONS

J. Woods

Forestry

1   Data from FAO Forest Products Yearbook, 1989, calculated solely from 1988 'roundwood production' figures.

2   It is assumed that 'roundwood' (which is synonymous with previously defined 'removals' in prior Yearbooks) is equivalent to 60 per cent of the total volume of wood actually cut (i.e. total cut equals 1.67 times roundwood production).

3   The 60 per cent figure is based on the amount of commercial stem wood which is available from the total tree above-ground biomass (see Hall and Overend (eds) (1987) *Biomass: Regenerable Energy*); thus only the stem and large branches are removed from the cutting site.

4   On a global basis, of the wood removed from site ('roundwood'), approximately half is used for 'industrial roundwood' and the remaining half for 'fuelwood + charcoal' (data from Yearbook). Historically, at least 50 per cent of the 'industrial roundwood' was 'lost' (predominantly as sawdust) during processing, most of which could be considered as potentially harvestable residues. The amount of wood which remains as residues varies widely and depends mainly on process efficiency and economics. More recently, and particularly in OECD countries, residues generated at the mill are at least partially used for other purposes, predominantly the manufacture of particleboard and MDF. Work is being carried out to quantify the impacts of the improvements in utilization of the wood residue fraction in mills.

5   'Potentially harvestable residues' include all on-site forestry residues ('r1', i.e. 40 per cent of total cut wood) plus all residues arising from 'industrial roundwood' processing at the timber mills ('r2', i.e. 50 per cent of 'industrial roundwood', calculated for each country) – see Figure 2.3. Practically, we assume that only 25 per cent of the 'potentially harvestable residues' are 'recoverable'.

Decisions as to the amount of residues that should be left on site for soil

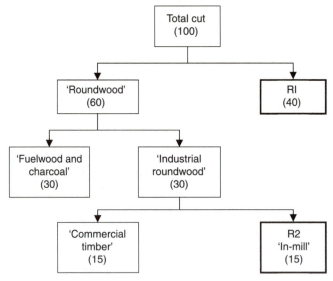

Notes: Total cut = 1.67*'roundwood'.
'Industrial roundwood' = 'roundwood' – 'fuelwood + charcoal'.
R1 = portion of total cut left on site, and is equivalent to 40 per cent of the total cut.
R2 = portion (50 per cent) of 'industrial roundwood' lost through processing.
'Potentially harvestable residues' = r1 + r2 = 55 (globally).
Forestry residues were analysed individually to deduce 'r1' and 'r2'.
Individual regional totals are aggregates of the individual countries.

Figure 2.3 *Diagrammatic breakdown of forestry production to show origin of residues (hypothetical example based on global averages)*

conditioning, nutrient recycling, etc., will vary greatly for any particular site. However, in our final calculation of biomass energy supplies by country or region, only 25 per cent of the 'potentially harvestable residues' are used to calculate the 'recoverable residues'. This should allow sufficient leeway for sustainable residue collection, but this can only ever be an informed decision made for each individual site, arrived at by taking fully into consideration all site-specific factors and preferably including ongoing monitoring.

Crop residues

1   Data from FAO Production Yearbook, 1989, for Cereals, Vegetables and Melons and Sugar Cane. Roots and Tubers and Sugar Beet data is from FAO's Agrostat database.
2   Since comprehensive global data is only available for crop production, the 'potentially harvestable residue' resource is estimated using residue production coefficients. These allow rough estimates to be made of the amounts of residues available per tonne of product; thus it is effectively a by-product-to-product ratio. For cereals, as an average for all types we have used a figure

of 1.3, that is, for every tonne of wheat, corn, barley, etc., grain harvested there is the potential to harvest 1300 kg (air dry) of residues ('potentially harvestable').

Sugar cane residues are also calculated from FAO production figures that provide the amount of cane stems harvested. Using the assumptions of Alexander (quoted in Carpentieri, Working Paper no 119, PU/CEES, Princeton, 1991) a residue coefficient of 0.55 was used to calculate the residue production figures (see Table 2.1). This figure is derived from:

— production of bagasse at 0.3 t (50 per cent moisture) per tonne of harvested stem;
— use of 'green tops and attached trash' (barbojo), at 0.25 t (50 per cent mc) per harvested tonne of stem.

The barbojo figure is lower than is used in Ogden et al (1990), (0.66 t/t cane stem). We assume that the 'detached leaves' will remain in the field to act as soil conditioner and are not a component of the collected barbojo.

3 We use a coefficient of 1.0 for vegetables and melons, which, although seemingly high, may be justified by a study which showed, in an extreme example, that only about 6–23 per cent of the harvested lettuce plant is finally eaten, depending on the season. There are other problems in dealing with factors such as changing water content, realistic levels of collectability for energy use and amounts used as animal feed. However, until better data is available, we will continue to use this deliberately simplistic coefficient. It should also be noted that vegetables + melons, roots + tubers, and sugar beet together only provide 3 per cent of the total 'potentially harvestable residues' while cereals provide 32 per cent and sugar cane 6 per cent of the total.

4 As with forestry residues, the amount of residues which can be removed from the field is determined by a wide variety of factors that are site dependant. However, we use only one-quarter of the calculated 'potentially harvestable

Table 2.1 *Crop residue production ratios*

Crop	Production coefficients		Energy content	
	t/t	Moisture	GJ/t (HHV)	Moisture
Cereals	1.3	Air dry	12	Air dry
Vegetables and melons	1.0	Air dry	6	Air dry
Roots and tubers	0.4	Air dry	6	Air dry
Sugar beet	0.3	Air dry	6	Air dry
Sugar cane	0.55	50%	16	Air dry

residues', and call these 'recoverable residues'. Thus, we allow for the worst cases where most of the residues will be required for protection against erosion, nutrient recycling and water retention.

5   The total energy content for 'potentially harvestable' crop residues (15.59 EJ developed, 21.51 EJ developing and a total of 37.10 EJ for the world) does not include estimates for residues from pulses, fruits and berries, oil crops, tree nuts, coffee, cocoa and tea, tobacco or fibre crops. Although regional estimates of their recoverable residue potential would be too small to appear in these tables they may still be a locally significant energy resource but will not often figure significantly on a national scale.

Dung

1   Dung production is calculated using data from FAO and UN Population Division (World Resources Institute). This data only shows the numbers of animals, disaggregated into different commercial species, and it is therefore necessary to use the dung production coefficients shown in Table 2.2 to estimate the total energy potential from animal dung production.

2   Due to the dispersed nature of dung, it is estimated that only 50 per cent of the dung actually produced is 'potentially harvestable'; we consider that only 25 per cent of the 'potentially harvestable' dung is 'recoverable', thus only one-eighth of the total dung actually produced by the animals is calculated as 'recoverable dung'. It should be noted that in a number of developing countries dung already plays a significant role in the domestic energy sector (e.g. India and China).

Table 2.2 *Dung production coefficients*

	Dung production rate: kg (oven dry) per day per animal	Energy content: GJ/t (oven dry)
Cattle	3.0	15.0
Sheep and goats	0.5	17.8
Pigs	0.6	17.0
Equines	1.5	14.9
Buffaloes and camels	4.0	14.9
Chickens	0.1	13.5

Note: The coefficients are predominantly based on developing country dung production rates and may significantly underestimate production rates in OECD countries.
Sources: derived from: Taylor, T. B. et al 'Worldwide data related to potentials for widescale use of renewable energy' Report no pu/cees 132, Center for Energy and Environmental Studies, Princeton University, NJ and Senelwa, K. and Hall, D. O. 1991 'A biomass energy flow chart for Kenya' (unpublished).

APPENDIX 2.2 DATA USED TO BUILD THE BIOMASS FLOW CHART FOR VANUATU

Table 2.3 *Summary of biomass energy production and use in Vanuatu*
(Forestry, Agriculture & Livestock)

VANUATU		Units	Annual average (2000–2003)
FORESTRY TOTAL WOOD CUT	Production (60% Stem Wood)	m	207,915
FORESTRY TOTAL WOOD CUT	Production (1.3t/m)	t	270,290
FORESTRY TOTAL WOOD CUT	Energy Content (15GJ/t)	GJ	4,054,343
% Total Biomass Energy Production			33
FORESTRY REMOVALS (USE)	Production	m	125,250
FORESTRY REMOVALS (USE)	Production (1.3t/m)	t	162,825
FORESTRY REMOVALS (USE)	Energy Content (15GJ/t)	GJ	2,442,375
% Total Biomass Energy Production			20
FORESTRY LOSSES	Production	m	82,665
FORESTRY LOSSES	Production (1.3t/m)	t	107,465
FORESTRY LOSSES	Energy Content (15GJ/t)	GJ	1,611,968
% Total Biomass Energy Production			13
LIVESTOCK	Stocks	Head	751,850
LIVESTOCK	Dung Production per Day	kg	388,990
LIVESTOCK	Annual Dung Production (365 days)	t	141,981
LIVESTOCK	Energy Content of Dung Produced	GJ	2,233,495
% Total Biomass Energy Production (Dung Prod)			18
LIVESTOCK	Dung Use	t	2,840
LIVESTOCK	Energy Content of Dung Used	GJ	44,670
% Total Biomass Energy Production (Dung Use)			0.4
LIVESTOCK LOSSES	Unused Dung	t	139,142
LIVESTOCK LOSSES	Energy Content of Unused Dung	GJ	2,188,825
% Total Biomass Energy Production			18
AGRICULTURAL CROPS	Area Harvested (Estimate)	ha	83,835
AGRICULTURAL CROPS	Crop Production (Food Use)	t	273,713
AGRICULTURAL CROPS	Residue Production	t	116,015

(Continued)

Table 2.3   (*Continued*)

VANUATU		Units	Annual average (2000–2003)
AGRICULTURAL CROPS	Crop Energy Content (Food Use)	GJ	4,562,228
% Total Biomass Energy Production (food)			38
AGRICULTURAL CROPS	Residue Energy Content	GJ	1,304,848
% Total Biomass Energy Production (residue)			11
AGRICULTURAL CROPS	Total Mass (Res + Prod)	t	389,728
AGRICULTURAL CROPS	Total Energy Content (Res + Prod)	GJ	5,867,075
% Total Biomass Energy Production (Res + Prod)			48
AGRICULTURAL CROPS	Residue Use	t	17,402
AGRICULTURAL CROPS	Energy Content of Residues Used	GJ	195,727
% Total Biomass Energy Production (Used)			2
AGRICULTURAL CROPS LOSSES	Residue + Crop (Mass)	t	98,613
AGRICULTURAL CROPS LOSSES	Residue + Crop (Energy Content)	GJ	1,109,120
% Total Biomass Energy Production (Losses)			9
TOTAL BIOMASS PRODUCED	Mass	t	801,998
TOTAL BIOMASS PRODUCED	Energy Content	GJ	12,154,913
% Total Biomass Energy Production			100
C CONTENT OF BIOMASS PRODUCED	Mass	t	639,732
TOTAL BIOMASS USE	Mass	t	456,779
TOTAL BIOMASS USE	Energy Content	GJ	7,245,000
% Total Biomass Energy Production			59
C CONTENT OF BIOMASS USED	Mass	t	381,316
TOTAL LOSSES	Mass	t	345,219
TOTAL LOSSES	Energy Content	GJ	4,909,913
% Total Biomass Energy Production			40
C CONTENT OF BIOMASS LOSSES	Mass	t	258,416

Source: Based on data from FAOSTAT (www.fao.org).

Table 2.4 *Agricultural biomass energy production and use in Vanuatu*

VANUATU		Units	Annual average (2000–2003)
Fruit*	Area Harvested	ha	1,635
Fruit*	Crop Production	t	20,638
Fruit*	Residue Production (Ratio of Prod : Res = 1.2)	t	24,765
Fruit*	Crop Energy Content (7GJ/t)	GJ	144,463
Fruit*	Residue Energy Content (9GJ/t)	GJ	222,885
Fruit*	Total Energy Content (Res + Prod)	GJ	367,348
Maize	Area Harvested	ha	1,300
Maize	Crop Production	t	700
Maize	Residue Production (Ratio of Prod : Res = 1.4)	t	980
Maize	Crop Energy Content (14.7GJ/t)	GJ	10,290
Maize	Residue Energy Content (13GJ/t)	GJ	12,740
Maize	Total Energy Content (Res + Prod)	GJ	23,030
Roots & tubers*	Area Harvested	ha	4,925
Roots & tubers*	Crop Production	t	39,750
Roots & tubers*	Residue Production (Ratio of Prod : Res = 0.4)	t	15,900
Roots & tubers*	Crop Energy Content (3.6GJ/t)	GJ	143,100
Roots & tubers*	Residue Energy Content (5.5GJ/t)	GJ	87,450
Roots & tubers*	Total Energy Content (Res + Prod)	GJ	230,550
Coconut	Area Harvested	ha	73,750
Coconut	Crop Production (nut)	t	210,250
Coconut	Residue (fibre) (Ratio Prod : Res = 0.33)	t	69,383
Coconut	Crop Energy Content (20GJ/t)	GJ	4,205,000
Coconut	Residue Energy Content (13GJ/t)	GJ	901,973
Coconut	Total Energy Content (Res + Prod)	GJ	5,106,973
Groundnuts	Area Harvested	ha	2,225
Groundnuts	Crop Production	t	2,375
Groundnuts	Residue Production (Ratio of Prod : Res = 2.1)	t	4,988
Groundnuts	Crop Energy Content (25GJ/t)	GJ	59,375
Groundnuts	Residue Energy Content (16GJ/t)	GJ	79,800
Groundnuts	Total Energy Content (Res + Prod)	GJ	139,175
TOTAL	Area Harvested (Estimate)	ha	83,835
TOTAL	Crop Production	t	273,713
TOTAL	Residue Production	t	116,015
TOTAL	Crop Energy Content	GJ	4,562,228
TOTAL	Residue Energy Content	GJ	1,304,848
TOTAL PRODUCTION	Mass (Res + Prod)	t	389,728
TOTAL PRODUCTION	Total Energy Content (Res + Prod)	GJ	5,867,075
TOTAL RESIDUE USE	Residue Use (15% of total)	t	17,402
TOTAL RESIDUE USE	Energy Content of Residues Used	GJ	195,727
TOTAL USE	Residue + Crop	t	291,115

(Continued)

Table 2.4 (*Continued*)

VANUATU		Units	Annual average (2000–2003)
TOTAL USE	Residue + Crop (Energy Content)	GJ	4,757,955
TOTAL LOSSES	Residue + Crop (Mass)	t	98,613
TOTAL LOSSES	Residue + Crop (Energy Content)	GJ	1,109,120

Notes:
Fruit*: bananas and other fruit.
Roots & tubers*: potatoes, sweet potatoes, cassava, taro, yams and others.
Source: Based on data from FAOSTAT (www.fao.org).

Table 2.5 *Livestock biomass energy production and use in Vanuatu*

VANUATU		Units	Annual average (2000–2003)
Cattle	Stocks	Head	135,250
Cattle	Dung Production per Day (1.8kg)	kg	243,450
Cattle	Annual Dung Production (365 days)	t	88,859
Cattle	Energy Content of Dung Produced (18.5GJ/t)	GJ	1,643,896
Pigs	Stocks	Head	62,000
Pigs	Dung Production per Day (0.8kg)	kg	49,600
Pigs	Annual Dung Production (365 days)	t	18,104
Pigs	Energy Content of Dung Produced (11.0GJ/t)	GJ	199,144
Horses	Stocks	Head	3,100
Horses	Dung Production per Day (3.0kg)	kg	9,300
Horses	Annual Dung Production (365 days)	t	3,395
Horses	Energy Content of Dung Produced (11.0GJ/t)	GJ	37,340
Goats	Stocks	Head	12,000
Goats	Dung Production per Day (0.4kg)	kg	4,800
Goats	Annual Dung Production (365 days)	t	1,752
Goats	Energy Content of Dung Produced (14.0GJ/t)	GJ	24,528
Chickens	Stocks	Head	340,000
Chickens	Dung Production per Day (0.06kg)	kg	2,040
Chickens	Annual Dung Production (365 days)	t	745

Chickens	Energy Content of Dung Produced (11.0GJ/t)	GJ	8,191
Human	Stocks	Head	199,500
Human	Dung Production per Day (0.4kg)	kg	79,800
Human	Annual Dung Production (365 days)	t	29,127
Human	Energy Content of Dung Produced (11.0GJ/t)	GJ	320,397
TOTAL PRODUCTION	Stocks	Head	751,850
TOTAL PRODUCTION	Dung Production per Day	kg	388,990
TOTAL PRODUCTION	Annual Dung Production (365 days)	t	141,981
TOTAL PRODUCTION	Energy Content of Dung Produced	GJ	2,233,495
TOTAL USE	Dung Use	t	2,840
TOTAL USE	Energy Content of Dung Used	GJ	44,670
TOTAL LOSSES	Unused Dung	t	139,142
TOTAL LOSSES	Energy Content of Unused Dung	GJ	2,188,825

Source: Based on data from FAOSTAT (www.fao.org); Hemstock (2005).

Appendix 2.3 Volume, density and moisture content

J. Woods and P. de Groot

Woody biomass, particularly fuelwood, production and consumption are normally measured by volume. But frequently in informal markets and household surveys the only record of fuelwood quantities produced, sold or consumed is a volume measure based on the outer dimensions of a loose stack or load containing air spaces between the pieces of biomass, such as the stere, cord, truckload, headload or bundle.

Two approaches can be employed to use such measures for energy analysis:

• to convert stacked volume to a weight, for example by weighing a number of samples with a spring balance (small load) or a weighbridge (truckload);
• to convert stacked volume to solid volume, for example by immersing loads (small ones) in water and then to measure the volume of water displaced.

Conversion of volume to weight

$$\text{weight (kg)} = \text{volume (m}^3\text{)} \times \text{density (kg/m}^3\text{)}$$

It should be noted that the density is the density of the biomass 'as received' and so in order to calculate the energy content it is also necessary to know the moisture content. The moisture content can also be estimated e.g. 'green' wood is typically around 50% moisture (wet basis; see below) and air dry wood is often around 15% moisture (wet basis), but this is highly dependent on the nature and duration of storage after harvesting. The moisture content also significantly affects the energy content (see below). The following sections describe the inter-relationships between density, volume, moisture content, gross calorific value (GCV) and net calorific value (NCV).

DENSITY

The density and moisture content are key determinants of the net energy content of a biomass feedstock. Specific gravity is often used to indicate the density of a substance. Specific gravity is the relative weight per unit volume of a substance when compared to water. The actual density of a substance is measured in units of kg per m^3 or g per litre. However, volume changes with temperature and so specific gravity data should be quoted at standard room temperature and pressure, or under stated conditions.

It is also important to know the physical state of the biomass for which the density is being stated. For example, stacked wood, chipped wood, wood pellets, loosely stacked straw, baled straw, etc. Density data on wood is often provided as stacked logs, or solid wood volume which is estimated by measuring the air voids in a wood stack. Standard energy densities usually refer to the solid biomass volume and not to the biomass as received unless directly measured.

If weight is used to determine the solid volume, density becomes an important factor since weight depends on density, and density varies within and between species. For example wood from young trees is less dense than from old trees of the same species, and sapwood is less dense than heartwood.

Various techniques are also used to make biomass feedstocks denser in order to reduce transport costs and to make them more manageable. For example, straw can be baled in-field increasing its density from c. 50 kg/m^3 up to 500 kg/m^3 depending on the type of baling system used. The density of wood can vary from as low as 150 kg/m^3 to over 600 kg/m^3 as stacked logs as shown in Table 2.6.

MOISTURE

The moisture content of biomass is often provided on a wet basis i.e. a 100 tonne consignment of wood is stated as having a 15% moisture content (wet basis) meaning that 15 tonnes of its mass is water and 85 tonnes oven dry biomass. To change from a dry to wet basis the following formula is used:

$$W = D/((1+D)/100)$$

Where D = moisture content as a percentage of the dry weight basis and W = moisture content as a percentage of the wet weight basis.

Table 2.6 *Example densities of biomass feedstocks*

		low	Density Kg per cubic metre (kg/m³) Inter-mediate	high
Sawdust		150	–	200
Wood chips		200	–	300
Logwood	30–50cm length	200	–	500
Logwood	100cm length	300	–	500
Pellets of sawdust or chips		400	500	600
Straw	Chopped straw		50	
Straw	High pressure bales	80	–	100
Straw	Big bales		100	
Straw	Pellets	300		500

Note: adapted from **Strehler & Stultze (1987)**

The reverse formula for changing from wet basis moisture content to dry basis moisture content is as follows:

$$D = W/((1 - W)/100)$$

If the moisture content is 100 percent on a dry basis then the wet basis moisture content = 50%.

ENERGY CONTENTS

The energy content of a biomass sample depends on its physico-chemical composition, primarily on the water and hydrogen contents. Its 'gross calorific' or 'higher heating' value is a measure of the energy content of biomass without any 'free' water. This completely dry biomass still contains chemically bound water and water that will arise as a result of chemical reactions during combustion. The GCV includes the latent heat of evaporation of this chemically bound water and in practice this energy can be recovered when using condensing combustion systems.

The 'net calorific' or 'lower heating' value of biomass is the energy content of biomass as received and excludes the latent energy that can be recovered via condensation.

Formulae are provided below for calculating both the GCV and NCV of a biomass sample.

GROSS CALORIFIC VALUE OR HIGHER HEATING VALUE OF BIOMASS

$GCV = 0.13491.X_C + 1.1783.X_H + 0.1005.X_S - 0.0151.X_N - 0.1034.X_O - 0.0211.X_{ash}$ [MJ/kg,d.b.]

Where:

X_C = %wt carbon content (dry basis)
X_H = %wt hydrogen content (dry basis)
X_S = %wt sulphur content (dry basis)
X_N = %wt nitrogen content (dry basis)
X_O = %wt nitrogen content (dry basis)
X_{ash} = %wt ash content (dry basis)

NET CALORIFIC VALUE OR LOWER HEATING VALUE OF BIOMASS

$$NCV = GCV \times (1 - (W/100)) - 2.447 \times (W/100) - 2.447 \times (H/100) \times 9.01 \times (1 - (W/100))$$

Where:

NCV = net calorific value (MJ/kg wet basis)
GCV – gross calorific value (MJ/kg dry basis; for wood this is typically 20 MJ per oven dry kg)
W = moisture content of the fuel in wt% (wet basis)
H = concentration of hydrogen e.g. wood biomass fuels c. 6.0%wt (dry basis); herbaceous biomass fuels c. 5.5%wt (dry basis)

The relationship between moisture content (wet basis), gross calorific value and net calorific value is shown in Table 2.7 below. This table shows that a biomass feedstock with a GCV of 21 MJ/kg and a moisture content of 70% would have an NCV of 4.19 MJ/kg as received.

Table 2.7 *NCV (red text; MJ/kg) as calculated from moisture content and GCV*

| | | Gross Calorific Value (MJ/kg) | | | | |
		21	20	19	18	17
Moisture	70%	4.19	3.89	3.59	3.29	2.99
(%wt wet basis)	60%	6.40	6.00	5.60	5.20	4.80
	50%	8.62	8.12	7.62	7.12	6.62
	40%	10.83	10.23	9.63	9.03	8.43
	30%	13.04	12.34	11.64	10.94	10.24
	20%	15.25	14.45	13.65	12.85	12.05
	10%	17.46	16.56	15.66	14.76	13.86

FURTHER READING:

Leach, G. and Gowan, M. Household Energy Handbook, An Interim Guide and Reference Manual. Washington, DC. 1987

Strehler and Stützle. Biomass Residues. In: *Biomass: regenerable energy*, edited by D. O. Hall and R. P. Overend, London: p. 75–102 John Wiley & Sons Ltd., 1997.

van Loo, S. and Koppejan, J. *Handbook of Biomass Combustion and Co-Firing*, Enschede:Twente University Press, 348 pages. ISBN 9036517737. 2002.

3

Assessment Methods for Woody Biomass Supply

Frank Rosillo-Calle, Peter de Groot and Sarah L. Hemstock

INTRODUCTION

Chapter 2 described general methodologies for biomass assessment. This chapter looks at the most important methods for accurately measuring the supply of woody biomass for energy, and in particular techniques for:

- forest mensuration;
- determining the weight and volume of trees;
- measuring the growing stock and yield of trees; and
- measuring the height and bark of trees.

and the energy available from:

- dedicated energy plantations;
- agro-industrial plantations;
- processed woody biomass (woody residues, charcoal).

REQUIREMENTS PRIOR TO AN ASSESSMENT OF BIOMASS SUPPLY

Assess the natural resources

An analysis of biomass energy resources requires accurate assessment of a country's natural resources. These should include:

- land type;
- vegetation type;
- soil composition;
- water availability;
- weather patterns.

Identify where you need to focus your assessment

Any supply assessment should be preceded by an examination of consumption in order to target the areas of greatest need and so determine where to direct the greatest efforts.

Be clear as to the purpose of the assessment

A biomass assessment is carried out with the intention to implement some kind of action. The nature of the action required will determine the nature of the assessment. It is therefore essential to define the objectives of the assessment you intend to carry out.

Identify your intended audience

The people you intend to read your report will determine the method you adopt to collect and present your data. As stated in Chapter 2, there are generally three types of audience for which assessment surveys are intended: policy makers, planners and project implementers. The type of information, and the manner in which it is presented, ranges from a generalized but succinctly presented picture for the policy maker, to detailed scientific data for the project officer. Your assessment may contain any level of precision of detail between these two extremes, or it may contain sections addressed to different audiences.

Decide whether a field survey is necessary

Field surveys are frequently used to generate much-needed data, but they are complex, time consuming, expensive and extremely demanding of skilled personnel. The problems are compounded in remote areas and harsh terrains. Other alternatives should be explored, such as analysis of existing data and collaboration with national surveys. Only consider mounting a field survey if there is no alternative, or if the areas involved are small. Box 3.1 illustrates a possible decision-making process when deciding the necessity for and detail of a field survey.

Box 3.1 A decision tree for a formal woodfuel survey

1   Is the problem defined, and scale and required precision established?

    If not, **stop!**

2   Check existing information.

    If adequate, **stop!**

3   Check, redefine and clarify problem.

    Repeat step 1.

4   After completion of steps 1–3, does existing knowledge need updating?

    If not, **stop!**

5   Is Rapid Rural Appraisal (RRA) more suitable?

    If 'yes', conduct RRA and complementary focused small surveys.

6   Is there sufficient information to stratify population?

    If not, go back to step 2; if still not consider **stop!** Otherwise prepare steps 1–5.

7   Estimate sample size and budget.

    If inadequate, repeat steps 1–5.

8   Are trained enumerators and data processors available?

    If not, could they be trained? If not, repeat steps 1–5.

    If 'yes', commence training/recruitment (concurrent with steps 9 and 10).

9   Design draft questionnaire.

10 Is pilot testing OK?

    If not, repeat steps 1–5.

11 If funds are available, staff trained, timing and seasonality OK, commence fieldwork and related supervision.

    If unsatisfactory, **stop or return to step 1.**

12 Data processing: checking, editing, coding entry and validation.

    If unsatisfactory, correct and go to step 13. If not corrected, **stop!**

13 Conduct statistical analysis and prepare tables of results.

14 Compare with initial hypothesis, previous results, local expert opinion and studies from other countries.

15 Return to step 1. Is further information still required?

    If not, submit report.

16 Conduct follow-up studies, RRA, debriefing of filed staff and spot surveys as required.

    Submit report.

Measuring biomass variation with agro-climatic zone and with time

Biomass yield varies with the:

- type of biomass and the species;
- agro-climatic region;
- management techniques employed in biomass production (for example, intensive or extensive forestry or agriculture, irrigation, the degree of mechanization, etc.).

These factors are important when making general estimates in the absence of detailed land use data. Furthermore, the productivity of biomass will almost certainly vary across seasons and over years. It is important to collect time-series data wherever possible. Particular care is needed if the biomass is not homogeneous.

Identify the scope and quality of existing data?

Much data is often already available, in the form of maps and reports, for example. Make a thorough search of the available literature before you begin your assessment.

Care is needed when using data from existing sources. Information on the availability of woody biomass is very patchy and there is a pressing need to improve the statistics on fuelwood, which constitutes the bulk of the wood used in many developing countries. Most published data on wood biomass considers only recorded fuelwood removal from forest reserves (although it is often not made clear that this is the case). The figures therefore ignore large areas of actual fuel supply, including:

- unrecorded removals from forests;
- trees on roadsides and community and farm lands;
- small diameter trees;
- shrubs and scrub in the forest;
- branch wood.

Remote sensing

Remote sensing (dealt with in detail in Chapter 6).

The assessment of land use

The method by which land use is assessed will depend on the level of information required. It is advisable to begin collecting data at the most general, aggregated level suited to policy-level analysis. Further enhancement and disaggregation will almost certainly be needed.

A proper analysis of agro-climatic zones will be necessary: if this information is not available, some form of remote sensing is the most viable option. The particular technique chosen will depend on local circumstances. At this stage, aerial photography and a full analysis of satellite imagery is usually too expensive in terms of both time and money.

An indication of the changes in land use patterns over time can provide a useful understanding of the evolution of local biomass resources, and enable predictions of likely resource availability in the future. It is therefore extremely valuable to examine the change in land use over time. This can be done in two complementary ways:

- the analysis of historical remotely sensed data if this is available;
- through detailed discussion in the field with the local population.

The accessibility of biomass

The availability and the potential supply of biomass are rarely the same. It is therefore very important to take into account the actual accessibility of biomass, although this is often very difficult to measure. There are three main physical and social constraints that restrict access to biomass:

- locational constraints;
- tenurial constraints;
- constraints derived from the land resource management system.

These are dealt with in detail in Chapter 2 under 'Assessment of accessibility'.

Key steps

An assessment of the supply of biomass involves the key points outlined below.

Assessment method
Decide which biomass assessment method is most suited to your needs. The method of estimating woody biomass, whether directly or indirectly, is determined by a number of factors, including the area in which it is growing, its variability and its physical size. You will probably have to put more effort into measuring scattered non-forest trees, because these non-traditional sources of wood are so neglected and show so much diversity of tree management.

Multiple uses of biomass
Consider the multiple uses of biomass, and the large quantities of biomass that may be available from different industrial processes such as sawmill waste, charcoal, etc.

Different types of biomass
Consider the different types of biomass resources, for example, woody biomass (firewood) and non-woody biomass (as agricultural residues).

Commercial value
Be aware of the commercial value of biomass. Bear in mind that biomass resources may be valued differently as an energy resource, depending on the local, regional or national tradition and culture. For example, animal residues can play a significant role in some countries such India, but are hardly used in others.

Fuel switching
Be aware of fuel switching. As living standards increase, or people move to rural centres, fuel preferences may change.

Secondary fuels
Be aware of the increasing importance of secondary fuels (e.g. biogas, ethanol, methanol, etc.). These fuels are obtained from raw biomass and are used in increasing quantities in modern applications, both in developed and developing countries.

BIOMASS ASSESSMENT METHODS FOR WOODY BIOMASS

The method for estimating woody biomass, whether directly or indirectly, is determined by a number of factors including:

- its type;
- the area in which it is growing;
- its spatial pattern;
- its variability;
- its physical size.

Woody biomass is perhaps one of the most difficult, but usually the most important, measurements to make. Every effort should be made to determine growing stock and the annual growth increment. The following is a summary of the steps involved.

Box 3.2 Multi-stage approach to woody biomass assessment

1   Review of existing data/maps for area
2   Low spatial resolution imagery –

The following are initial questions that may be helpful in formulating your assessment strategy for woody biomass (based on ETC Foundation, 1990, Box 12 p33).

What types of woody vegetation are present in the area?
- Forests: planted or natural; main species;
- Bush land;
- Open woodland;
- Trees in and around farming areas: woodlots, windrows, scattered trees in cropland, trees on compound;
- Trees in public places: markets, roadsides, along canals.

What is the condition of these vegetation types?
- Well maintained or neglected;
- Gaps because of heavy cutting;
- Natural regeneration;
- Pruning, pollarding;
- Collection of dead wood;
- Fresh stumps;
- Coppices;
- Litter;
- Erosion.

Do you observe any transportation or trading of forest or tree products?
- Heaps of wood on the roadsides;
- People transporting wood, charcoal, fruits, tree leaves, etc.;
- People selling wood, charcoal, fruits, bark, roots, medicines, etc. in markets or elsewhere.

Do you see any activity related to processing or utilization of tree products?
- Sawing or splitting;
- Burning charcoal;
- Fencing;
- Building;
- Processing of fruit;

- Basket making;
- Feeding leaves to cattle, etc.

Do you observe any activity related to tree regeneration and management?
- Tree nurseries;
- Transportation;
- Selling of seedlings;
- Young trees or newly planted cuttings;
- Pruning, clipping, thinning, clearing, coppicing.

Wood is rarely grown specifically for fuel, because in most cases woodfuel is obtained at zero or near zero cost. Wood sold for purposes other than fuel has a much higher market value, and consequently normally has priority over use of wood as fuel. It is therefore important to consider the non-fuel uses for wood.

However, considerable waste is generated when trees are converted to wood products. In the forest or on farm the buyer (or seller) only removes logs of specific dimensions; branchwood and crooked stem wood – which might amount to 15–30 per cent or more of the above-ground volume – may be left when the trees are felled. Sawn wood may only account for between one-third and one-half of the original sawlog, leaving the other half to two-thirds as waste. Even the conversion of sawn wood into poles or timber produces more waste. All these waste materials are potentially burnable, especially if they are near to the demand, and must therefore be included in the assessment.

Appendix 3.1 provides more details on various methods of projecting supply and demand. You will probably have to put more effort into measuring scattered non-forest trees, because these important sources of wood are often neglected and involve considerable diversity in tree management.

Techniques are available for measuring tree biomass by volume and by weight. For the reasons stated already in Chapter 2, weight is the most suitable measurement for fuelwood surveys. However, because the forest industry measures timber by volume, the techniques for determining the volume of tree stems and larger branches are far more developed. The techniques for assessing both weight and volume are outlined here. However, to enable comparison, all measurements should be converted to metric weight units.

It is an easy matter to find the total weight of the tree, including the crown, if tables exist that give the relationship between stem and branch volume for the species being assessed. As there are standard techniques for calculating the volume of stems and crowns, it may be easier where tables are available to first calculate the volume, and use this to estimate the weight.

Satellite imaging and/or aerial photography are useful tools to estimate woody biomass. As branch volume and weight can vary from 10 per cent for trees grown in uniform plantations to over 30 per cent for free-standing trees, it is essential to augment imaging data with field measurements. A literature search may also provide useful information. A combination of information from these sources should provide accurate estimates of biomass volume and weight estimates for individual sites.

Where twigs and leaves are collected for fuel, it is necessary to carry out destructive sampling of a small number of trees to provide measurements of the leaf, branch, stem and root weights. You can then estimate the total available tree biomass per unit area. Appendix 3.2 gives further details of how to measure fuelwood resources and supply.

Forest mensuration

Mensuration – the measurement of length, mass and time – incorporates principles and practices perfected by land surveyors, foresters and cartographers. Forest mensuration is the tool that provides data on forest crops, or individual trees, felled timber, and so on. The principles of measuring trees are given in several books (for example, see Husch et al, 2003), but there is only one satisfactory way to measure shrubs and hedges and that is to cut them down, weigh them and obtain a relationship between volume of the whole shrub, including air space, and weight.

There are many techniques used by commercial forestry to measure forest, individual tree parameters, branches, bark, volume, weight, etc. that you can borrow to estimate total or partial biomass availability.

Individual trees should be described quantitatively by various measurements, or parameters, the commonest of which are:

- age;
- diameter of stem – over or under bark;
- cross-sectional area – calculated from diameter of stem;
- length or height;
- form or shape – trees are not cylindrical;
- taper or the rate of change of diameter with length;
- volume over or under bark; volume may be calculated to varying top diameters;
- crown width – a parameter that can be measured both in the field and/or via aerial photographs;
- wood density.

Measurements of tree crops, woodlands, plantations and forests require further measurements including:

- area – surveyed or estimated from maps or aerial photographs;
- crop structure – in terms of species, age and diameter;
- total basal area per hectare;

- total biomass, dry weight per hectare;
- total energy resource per hectare.

For even-aged, uniform plantations of a single species the following measurements are also frequently used:

- average volume per tree;
- average stem basal area per tree;
- diameter of tree of average basal area;
- average height, per tree (King et al, 1990).

Traditionally, foresters measure trees by stem volume or by weight, which includes the moisture content. It is possible to estimate the stem volume of a tree with height and diameter measurements, and the per hectare volume with average height and basal area measurements. The unit weight of these trees can be calculated if the density of the wood is known.

Field surveys providing on-the-ground biomass data have not been done, partly due to lack of interest in biomass, and partly because of the lack of appropriate calibration curves relating individual tree or bush biomass to easily measurable tree or bush dimensions. Extensive mensuration data is available for the very limited number of tree species commonly cultivated in plantations. When only a few species are present, it may be possible to estimate the woody biomass of plantations using land use data combined with biomass tables that can provide the volume or weight of biomass for a given species. These tables are particularly useful where the vegetation is fairly uniform. At present very few biomass tables are available. Once drawn up, however, such tables allow the subsequent survey to be carried out rapidly.

However, natural forests in subtropical environments are composed of a multitude of differing trees on which no information is available. The standing biomass of trees in natural tropical forests has been handled on a case-by-case basis (see Allen, 2004). Surveys have produced regression equations against stem diameter, stem circumference, stem basal area, tree crown dimensions and combinations of these on a case-by-case basis, resulting in many custom-made regressions, suiting the aims of the various researchers (Tietema, 1993).

For example, Tietema (1993) carried out studies in Botswana aimed at providing a set of regression equations relating external tree dimensions to total above-ground biomass. This was done by measuring the height and the diameter of the crown, together with the diameter of the stem at 'ankle height' (approximately 10 cm above ground level) of sample of trees. The trees were then cut down so that the weight of the total fresh biomass could be measured. For multi-stemmed trees the stem basal area and the tree weight of the individual stems were determined. The results show that in the regression of stem basal area against weight, there is a great similarity between the regression lines of a range of tree species in Botswana and also in Kenya with three different species.

Thus, in general, the set of regression lines offers a realistic and flexible possibility of carrying out extensive on-the- ground surveys of tree standing biomass. This flexibility, according to Tietema is very important in determining the effect of wood harvesting, establishing tree stock and the mean annual increment (MAI), and in assisting the interpretation of remote sensing data. However, it may not always be feasible to use destructive sampling as part of the survey, for example, when trees are incorporated in agricultural areas.

THE MAIN TECHNIQUES/METHODS FOR MEASURING WOODY BIOMASS

This handbook considers in detail only the main techniques that can be applied to assess woody biomass for energy.

Determining the weight of trees

It is possible to determine the weight of specific tree species in natural environments using measurements of the stem diameter (breast height, 1.3 m, or 0.4 m), stem circumference, stem basal area (0.1 m), tree crown dimensions, and combinations of the tree and crown dimensions. Due to the differing nature of the linear regressions obtained, making a comparison between species is very difficult. Figure 3.1 is derived from destructive measurements for British native woodland species.

As mentioned above, foresters are mostly interested in stem volumes, which give a considerable underestimate of the wood available for fuel. For example, branches and roots, which are important sources of fuel, make up well over 50 per cent of the tree wood. Where wood is scarce, twigs and leaves are also collected for burning. When assessing biomass for energy purposes, your calculation should include leaves, twigs, branches, stems and roots. Note that tree roots can amount to some 30–40 per cent of total woody biomass production, and up to approximately 55 per cent of above-ground woody biomass production. However, unless there is a change of land use, or a severe shortage of fuel, tree roots are generally not burned, although this does not apply to roots of shrubs and bushes.

Destructive sampling of a small number of trees to provide sample measurements of the leaf, branch, stem and root volumes or weights will allow estimates of the available biomass per unit area to be made.

Once the regression for a species or group of woody biomass species has been established, the most useful and easiest-to-measure regression is probably that between stem basal area and tree weight. This technique requires only

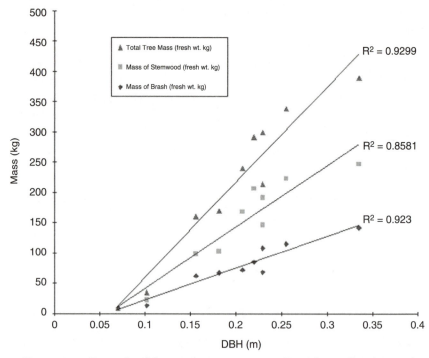

Figure 3.1 *Example of destructive measurement (British woodland species)*

two measurements of the diameter for each stem (single or multi-stemmed trees).

The technique is straightforward and requires only simple equipment, for example, callipers, scales and drying apparatus, for establishing the regression. Thereafter, only callipers are required for rapid measurement of the necessary sample of trees. Establishing the best sampling technique for small or large areas is probably as important as the actual measurements themselves.

To obtain weight estimates, a sample of trees has to be felled to determine the green, air-dry and possibly oven-dry density (weight/unit volume). While wood loses weight when it dries, there is little shrinkage until it gets down to about 10 per cent moisture content, therefore knowing volume and average density, the weight (air dry or oven dry) can be determined with sufficient accuracy.

Attempts have been made in the past decade to estimate the amount of woody biomass available using remote sensing techniques but the results of these studies were generally too inaccurate to be used for planning purposes. This has partly been caused by a lack of clear understanding of the relation between on-the-ground biomass and spectral reflection measured by the various satellites. However, important advances have been made in recent years that allow more accurate measurements at lower cost (see Chapter 6).

Techniques for measuring the volume of trees

Traditionally, to measure the volume of trees, foresters use:

- diameter at breast height (DBH);
- total height and crown measurements (diameter plus depth) to estimate individual tree volumes;
- mean height, basal area at breast height and mean crown measurements.

Below are a few examples of volume measurement used by the commercial forestry industry most of which can be done through specific computer programs (see www.woodlander.co.uk/woodland/, from which the following examples have been taken).

Volume measurement of trees and roundwood

The volumes of trees and roundwood are commonly measured both for the purposes of management and for their sale. 'Roundwood' means the products that are produced when a tree is crosscut into lengths. These products may be sawlogs, pulpwood, fencing or other material. The unit of volume is the cubic metre. Volume may be measured in different ways, depending on what is being measured. This is because the practicability and cost of a particular measurement method may depend on the product. The measurement of volume usually requires a measurement of diameter and of length.

Diameter   Diameter is measured in centimetres. After a diameter measurement is taken the diameter obtained is rounded down to the nearest centimetre. For example, diameters measured as 14. 9 cm or 14.4 cm. are rounded down to 14 cm.

Length   Length is measured in metres and rounded down. Lengths under 10 m are rounded down to the nearest 0.1 metre, while lengths over 10 m are rounded down to the nearest metre. Thus, lengths of 16.75 m and 18.3 m would be rounded down to 16 m and 18 m, respectively. Lengths of 6.39 m and 3.95 m would be rounded down to 6.3 m and 3.9 m.

When measuring the length of a felled tree, the length is usually measured to a top diameter of 7 cm. The different methods of volume measurement are as detailed below.

- The 'length and mid-diameter' method – this is used for felled trees and for situations where it is reasonably easy to get to the point at which the mid-diameter has to be measured.
- The top diameter and length method for sawlogs – special conventions apply where reasonably uniform logs are being sold in quantities of at least a lorryload. Their volume can be estimated from top diameter and length, making an assumption on taper.

- The top diameter and length method, for small roundwood: Volume may also be measured from top diameter, again making an assumption on taper.

Mid-diameter and length volume measurement method   The volume is dependent on the length of the tree or roundwood piece and the diameter at the middle of the rounded-down length. Length and diameter are, of course, measured according to the conventions stated above. Examples of the 'mid-diameter' and length method of measuring volume are given below.

- Roundwood less than 20 m in length:
  — measured length: 4.45 m, rounded down to 4.4 m;
  — measured mid-diameter (at 2.2 m): 14.6 cm, rounded down to 14 cm;
  — from the tables (see website www.woodlander.co.uk/woodland/voldfrm.htm#conventions), the volume for a length of 4.4 m and 14 cm diameter is: 0.068 m^3.

- Measuring the volume of a felled tree by this method:
  — if the length to 7 cm top diameter is 20 m or less, then the volume of the tree is measured according to the procedure outlined above;
  — if the length of the tree to 7 cm top diameter is greater than 20 m then the convention is to measure the volume in two lengths. To obtain the two lengths, divide the tree length by two. The first (butt length) will be the rounded down length of the bottom half of the two halves. The second length will be the remaining length. For example:

  > tree length = 37 m
  > tree length/2 = 18.5 m
  > first (butt) length = 18 m (mean diameter taken at 9 m)
  > second length = 19 m (mean diameter taken at 9.5 m along the length).

The volume of each length is obtained from the mid-diameter/length table and the totalled volume is the volume of the tree (see www.woodlander.co.uk/woodland/).

If local volume tables are available for the tree species being assessed, it is possible to use the diameter at breast height (DBH) alone to determine both the volume of the stem and total tree volume for single species stands. DBH is widely used as an independent variable in estimating total wood weight in a tree, but this is applicable only to well defined species and mostly to plantation and commercial forestry. However, even within the same species there can be a substantial variation in volume for trees of the same DBH. Therefore, general volume tables should be used that take height into consideration as well. Appendix 3.3 provides further details of how to measure tree volume.

Foresters have devised formulae to calculate weight from volume, so that it is an easy matter to find the total weight of the tree, including the crown, from tables that give the relationship between stem and branch volume. How useful these techniques can be for measuring fuelwood is not always clear. Furthermore, there are also standard techniques for calculating the volume of stems and crowns. It may therefore be easier to first calculate the volume and then use this to estimate the weight.

So if the average height and the tree taper are known for a particular species it is possible to estimate both the volume of the stem and the total volume of the tree above the ground.

The simplest way to calculate the volume of a tree is by using the formula:

$$v = \pi r^2\, h \times f$$

where:

v is the volume of the tree
r is the radius at breast height
h is the total height and
f is the reduction factor to allow for the taper on the tree.

The reduction factor (f) can range from 0.3 to 0.7 and has to be calculated from felling a number of trees and measuring individual logs.

Similarly, it is also possible to estimate the crown wood volume from the stem volume. A number of trees are felled to obtain the ratio of wood in the stem to that in the crown. This calculation is then used to estimate crown volume in the remaining trees from stem volume alone.

In closed stands, the per hectare stem crop volume (V) is given by the formula:

$$V = G \times H \times F$$

where:

G is the mean basal area per hectare
H is the mean height and
F is the mean reduction factor.

The basal area (G) is easily determined using an angle count measure (or rela-scope technique). A hypsometer can be used to obtain the height of a number of trees to determine mean height (H). The principles of measuring trees are given in several books (e.g. Philip, 1983; Husch et al, 2003).

There is only one satisfactory way to measure shrubs and hedges and that is to cut them down, weigh them and obtain a relationship between volume of the whole shrub including air space and weight.

Measuring growing stock of trees for fuel

All the above give a measure of growing stock, but increment or annual yield is required as well. If it is possible to calculate the total growing stock using the above techniques, it is then possible to estimate the annual increment. Note that you cannot assess annual increment by simply dividing growing stock by the rotation age, because trees do not put on equal annual quantities of biomass throughout their lifetime.

Some fuelwood crops may have rotations as short as one or two years, whereas in natural forest the rotation may be in excess of 100 years. Natural forest has a very large growing stock, but a relatively low annual yield, of the order of 2 m³ to 7 m³ per ha. Table 3.1 summarizes the relationship between age, height and growing stocks based on data from British woodlands, as does Figure 3.2. Better management can reduce the rotation and increase the annual yield so that there is a quicker turnover of stock. You should also make a thorough literature search of work on whole tree measurements and biomass grown on various sites and under different rainfall regimes (Hemstock, 1999).

Allometric, or dimensional, analysis is another option for measuring total tree biomass. Allometry is an old and widely used technique in estimating timber volume and weight for commercial species. However, variations in scaling arise in trees of different shapes and densities, and also in trees that have been extensively managed by pruning, lopping, etc. (e.g. farm trees). Thus, there is no

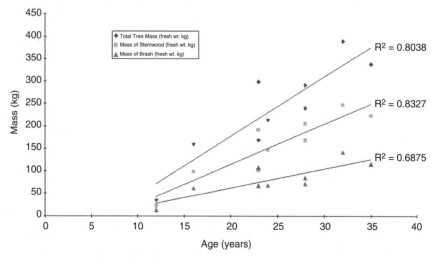

Figure 3.2 *Relationship between mass and age of trees in British woodlands*

Table 3.1 *Summary of broadleaf woodland stocking from UK data*

Species/woodland description	Average age (yr)	Average height (m)	Stocking (trees ha⁻¹)	Reference
Farm woodland (unmanaged)				Hemstock, 1999
Silver birch	26	11	463	
Oak	28	11	396	
Alder	23	10	261	
Beech	14	9	39	
Ash	15	9	10	
Total	**25**	**11**	**1168**	
Ash grown for timber (managed)	26	14	578	Forestry Commission (1998)
	44	16	437	
	69	22	248	
Sycamore grown for timber (managed)	27	15	604	Forestry Commission (1998)
	40	19	302	
	70	26	278	
Alder managed for timber (managed)	15	15	1247	Forestry Commission (1998)
	27	18	261	
	40	19	226	
Beech managed for timber (managed)	23	12	1161	Forestry Commission (1998)
	37	14	1156	
	46	17	889	
	66	18	631	
	97	21	307	
Oak managed for timber (managed)	37	15	707	Forestry Commission (1998)
	46	17	498	
	77	21	236	
	97	25	146	
Broadleaf stand grown for timber				Blyth et al (1987)
Establishment	0–10	0–5	1000–3000+	
Thicket	5–20	2–10	1000–2000	
Early thinning	15–50	8–18	500–1000	
Late thinning	30–100+	15–30+	150–600	
Mature	40–150+	18–30+	70–300	

Source: Hemstock (1999).

single universal formula for measurement applicable to all trees. Conifers differ from hardwoods, tropical woodland (savannas) trees from tropical forest trees and so on. Methodologies should be chosen according to the particular situations and circumstances. For forest species, especially tropical forest species, selected trees are usually measured along line plots at specific intervals using such measures as diameter (DBH), height and crown measurements. Trees outside the forest are measured in a similar fashion. However, these measurements tell us nothing about the area covered by the different woody types. For dense formations, area data can be obtained from satellite pictures and is applicable to most countries (Allen, 2004).

The most accurate way to obtain an estimate of density is to use satellite or aerial photographs. Once these data are known, biomass volume and weight estimates can be made for individual sites.

Satellite photography has great potential as it offers the possibility of covering large areas with a high degree of definition. But the technique still needs improvement. If the trees are scattered and sufficient resolution is possible, the assessment of height and crown dimensions could be used to estimate DBH of the trees. If the woodlands are relatively undisturbed, and contain a cross-section of most ages and stem diameters, the approximate volumes and/ or weight may be estimated from crown cover. Of course, some field work is necessary to obtain this volume relationship. Crown cover may also be estimated directly from ground measurements and from aerial observations (see Chapter 6).

Measuring annual yield of trees

Biomass production depends on the quality of the growing site, rainfall, tree species, the planting density, the rotation age and management techniques. Wider spacing produces larger trees, and planting more trees per hectare will not increase production once there is canopy closure. However, biomass production per unit area is more or less constant for trees of the same age on similar sites for a fairly wide range of planting densities.

You need to obtain the CAI (the increase in volume of a tree during a year) and the average CAI over several years, that is, the MAI. Thus:

CAI = volume increase per tree per year

MAI = volume increase per tree of a stated number of years.

While satellite photography may be valuable in determining areas of dense woody biomass, it cannot provide accurate data on either growing stock or annual increment. There is, at present, no substitute for ground measurements. Once the total growing stock has been calculated, it is possible to estimate the annual increment. The CAI is estimated by dividing the growing stock by half the rotation, that is:

$$I = GS \times R/2$$

where:

I = annual increment
GS = growing stock
R = rotation age.

This formula works well, except where the rotation age is small (for crops with a one year rotation GS = 1 and not 2GS). Note that you cannot assess annual increment simply by dividing growing stock by the rotation age, because trees do not put on equal annual quantities of biomass throughout their lifetime.

Yield of wood per unit area

The next step is to determine the weight or volume of wood per unit area. To calculate the sustained yields of wood and residues it is necessary to know:

- growth rates (foresters measure volume increments by taking periodic measurements of stem diameter, height, etc.);
- reproductive rates;
- rates of loss (from tree death and harvesting);
- quantities of wood harvested commercially;
- quantities and spatial distribution of non-commercial fuelwood collected;
- quantities of residues produced, and the amount it is necessary to return to the soil for sustained tree growth.

To calculate the annual increment, it is best to measure trees at periodic intervals, making sure that all removals from the trees are accounted for. One method is to establish permanent or temporary sample plots in fields or forest sites. The management techniques should also be carefully noted as this greatly affects the quantity and quality of the wood produced.

Once established, the scope and number of sample plots should be enlarged to get more accurate measures of at least all above-ground woody materials. As an alternative, trees at various stages of development could be measured and used to plot growth over time. Sample plots for trees and shrubs outside the forest are necessary in order to obtain better removal statistics from this important source of supply. For this technique, it is important to ensure that the agency responsible for the project has the necessary infrastructure to carry out this task. Table 3.2 illustrates standing stocks (stemwood and whole trees), MAI and woodland age, from British broadleaf forests.

Alternatively, yield tables are available which predict the growth and yearly yield of single species stands. The rotation or lifetime is defined as the point of maximum mean annual increment (MAI). Yield tables and volume tables are usually available for single plantation species for stem volume only, but can be adapted to give total volume by determining the relationship between stemwood and total wood.

Table 3.2 *Summary of broadleaf woodland production estimates from UK data*

Woodland description	Standing stock (t ha⁻¹)[a] (stemwood)	Standing stock (t ha⁻¹)[a] (trees)	MAI (t ha⁻¹)[a] (stemwood)[b]	MAI (t ha⁻¹)[a] (trees)	Average age of woodland (yr)	Reference
Predominantly beech (thinned)	58		3.4		17	Forestry Commission (1998)
	263		4.8		55	
	260		3.3		80	
Oak (thinned)	82		2.2		37	"
	245		3.2		76	
	329		3.0		110	
Ash (thinned)	85		2.4		35	"
	157		2.4		66	
	329		3.0		110	
Sycamore (thinned)	167		3.7		45	"
	217		3.8		57	
Alder (thinned)	98		4.4		22	"
	119		3.4		35	
	229		4.1		56	
Hornbeam (thinned)	124		3.8		33	"
	155		3.2		48	
	225		3.4		66	
Mixed broadleaf	150–200		3–4		50	Poole (1998)
Oak	80		1.6		50	
Ash/Syc/Horse Chestnut	108		3.6		30	
Ash/Sycamore/ Oak	110		4.0		28	
Oak/Sycamore (farm woodland)	152		3.0–3.4		48	
Mixed broadleaf (7 cm diameter and over)	170–180		5		35	Prior (1998)
Predominantly beech		230		5	46	Hemstock (1999)
Silver Birch/ Oak/Alder/ Beech (farm woodland)	173	260	6	9	30–40	
Poplar[c] (4.6 m spacing unthinned)	137		11.4		12	Forestry Commission (1988)
	420		19.2		22	
	687		21.4		32	
	857		20.4		42	
	957		18.8		52	
Poplar[c] (7.3 m spacing unthinned)	78		6.5		12	"
	260		11.7		22	
	441		13.8		32	
	569		13.5		42	
	648		12.4		52	

Sweet chestnut coppice		25		2.5	10	Forestry Commission (1988)
		160		8	20	
		405		13.5	30	
High forest[d] – mixed broadleaf	25		5.0		5	Forestry Commission (1984)
	148		4.9		30	
Sweet chestnut coppice[e]		11		2.2	5	Ford and Newbould (1970)
		38		3.5	11	
		85		4.7	18	

Notes:
[a] Assumes fresh weight production unless otherwise stated.
[b] Stemwood = removals from site of harvest (usually minimum diameter of 7 cm, consisting of wood from the base of the tree to the point in the crown where the trunk is no longer visible and including the bark).
[c] Conversion factor (m^3:t) = 1.0:0.99 (Forestry Commission, 1988).
[d] High forest is woodland in which the main purpose is to grow utilizable timber.
[e] Dry weight production.
Source: See Hemstock (1999).

Measuring tree height

Tree height is important since it is often one of the few variables used in the estimation of tree volume and thus it is a common attribute measured in forest inventories. 'Tree height' can have several meanings and can lead to practical problems (see Appendix I Glossary of Terms under 'Tree height measurement' for definitions).

To be able to measure the height of a tree, the top must be visible from a point from which it is feasible to see most of the stem bole. The methods used generally depend on the size of the tree to be measured. Direct methods (rarely used) involve climbing or using height measuring rods, and indirect methods (commonly used) use geometric and trigonometric principles using a theodolite or clinometer (see Figure 3.3). Readings for θ (angle to top of tree) and δ (angle to base of crown) are obtained from using a theodolite or clinometer and cos and tan values can be found in appropriate mathematical tables.

Measuring tree bark

Bark is the outer sheath of the tree. Some tree species shed bark while others have persistent bark which is removed when trees are cut. We should not be too concerned with the various techniques used to measure bark in standing trees, but should be aware of its importance as it is widely used for fuel from felled trees. Knowing how much bark a forest or plantation can generate is useful.

Knowing bark volume is particularly important in commercial logging, both when bark is a valuable product that can be sold separately, or when it is left in the ground to rot or be collected as fuelwood. On felled trees bark thickness is measured directly at the cut ends. You should be aware of the various techniques used to estimate total bark volume if you need to be able to calculate bark

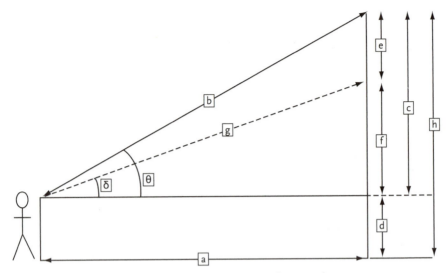

Figure 3.3 *Height measurement and crown dimensions*

Notes:
a = distance of theodolite from tree.
b = a/cos θ.
c = a tan θ.
d = height of theodolite.
e = crown height = c − f.
f = a tan δ.
g = a/cos δ.
h = height of tree = c + d.
θ = angle to top of tree.
δ = angle to base of crown.

volume of logging as a source of energy (see http://sres.anu.edu.au/associated/mensuration/bark.htm).

To measure bark thickness:

- place the bark gauge perpendicularly against the tree and push the arm until the whole of the bark has been traversed;
- read the bark thickness on the scale;
- take another measurement at a point diametrically opposite the first point of measurement;
- take the arithmetic mean.

METHODOLOGY FOR ASSESSING ENERGY POTENTIAL FROM FORESTRY/CROP PLANTATIONS

Despite the expected increased role of dedicated forestry/crops energy plantations (see Chapter 1), currently there is little commercial experience with such plantations. The only country which has large-scale plantations is Brazil where there are some 2.5 Mha of eucalyptus used in the industrial production of charcoal for steel, metallurgy, cement, etc. Sweden has about 14,000 ha of willows intended for power generation, which currently are used for other non-energy purposes. The USA has approximately 50,000 ha and many other countries also have energy plantations, but nothing on a grand scale and mostly for non-energy purposes, despite energy generation being the original idea.

A brief summary of the methodology for assessing energy potential from dedicated plantations follows. For further details reference should be made to Kartha et al (2005) pp104–118.

Assessing the potential resources from dedicated energy plantations should be much simpler than dealing with residues and forests (natural, on-farm, commercial plantations for non-energy use, etc.) for various reasons:

- there are fewer parameters to consider (i.e. energy is the main purpose rather than the source being multi-purpose, as is often the case when energy is just a by-product); the basic concerns being harvested area, yield/ha, energy content, etc., using conventional forestry or agricultural criteria;
- dedicated energy plantations respond to specific parameters required in modern applications such as clear financial criteria which are widely understood (i.e. is there a market and if so is it cost competitive, or can the land be put to better alternatives uses?).

The methodology for assessing dedicated energy plantations can be used at any scale (regional, national, village level) and at any desired level of disaggregation and type of energy crop. Kartha el al (2005) have identified six major steps for wood energy which are summarized below.

Step 1: Estimate areas of potentially available and suitable land

Land is the key factor and you therefore need to classify land into various categories: land availability and land suitability for energy plantations. This is not a simple as it may first appear, given the multiple uses of land, particularly in relation to food production, which should be given a higher priority.
Land *not available* includes:

- legally protected land (e.g. national parks, forest);
- socially protected land (e.g. amenity areas, forests, woodlands);
- built-over land (e.g. urban, industrial areas).

Land *not suitable*:

- climatic constraints (e.g. low precipitation, high temperature);

- terrain and soil constraints (e.g. steep slopes, rocky soil);
- remoteness and lack of infrastructure (e.g. difficult access, weak local energy demand);
- productive land under high-value crops (e.g. high-value timber plantations).

Step 2: Estimate yields associated with areas of suitable land

After assessing land availability and suitability, the next step is to find the yields for which standard production practices can be used. Most of the agricultural and forestry production models can be used to estimate yields.

Step 3: Calculate the gross potential resource (by land type and/or sub-regions)

This step merely involves multiplying the area of available and suitable land (step 1) by associated yields (step 2) to produce an initial estimate of the gross potential resource from the energy crop in question, both in total and by sub-region or land type, depending on the level of land disaggregation used in step 1 (see Kartha et al, 2005).

Step 4: Estimated production costs and delivered energy prices

The costs of growing biomass for energy are similar to those of agriculture and forestry, although with some differences. Using the terminology of forestry (Kartha et al, 2005), the basic cost categories are:

- the *stumpage* cost (the capital cost of establishing the crop and maintaining the crops until it is harvested);
- harvesting and haulage to the fuel preparation site;
- fuel preparation (e.g. chipping and drying);
- transportation (to the market or power plant);
- overheads and fixed costs.

Step 5: Estimate delivered energy prices (and ranges)

Once the cost of all the items identified in the previous steps have been established, it is not difficult to calculate the cost of energy generated and associated sale prices. The best way is to use a computer spreadsheet to tabulate all costs and revenue for each year (e.g. number of harvest rotations), discount these at chosen interest rate(s) and calculate measures of financial worth for hypothetical plantation projects (e.g. benefit/cost ratio, net present value (NPV) and internal rate of return (IRR)).

Step 6: Final integration and assessment of energy crop potential

The standardized results from step 5 can be used in three main ways to arrive at the final analysis of the total energy potential and associated prices:

1   For each discount rate and assumed net margins, locations can be filtered out as being suitable for energy crops. Rejections can be based on very high breakeven costs relative to competing energy prices.
2   For each discount rate and assumed net margin, the same process can be used to tabulate and plot energy outputs against price to produce a cost–supply curve.
3   If the data are available on how crop yields respond to site quality and greater production inputs and costs, the interactive process is outlined in section 1.3 (see Kartha et al (2005) pp17ff). If the data are available on how crop yields respond to site quality and greater production inputs and costs, the interactive processes are outlined in Chapter 1, 'Biomass Potential' and Appendix 1.1 can be applied to provide a broad picture of energy crop assessment.

Agro-industrial plantations

These crops are extremely diverse and no one method is suitable for determining the standing volume or weight and yield for all crops. First, the information already available should be obtained by carrying out a literature search. A field survey may then be necessary.

On-farm trees

Unfortunately, few reliable data exist concerning on-farm trees, which are extremely important as a fuelwood source. A field survey is the only way to collect accurate, detailed data for on farm trees. Farm trees are managed in many ways, and some diameter classes and ages are usually absent in any one sample. Aerial photographs are useful for estimating percentage cover and basic tree measurements, but this can only ever be supplemental to a ground survey. If specific sample plots are chosen, then average volume estimates for above-ground biomass per unit area can be determined. If the density of the wood from the particular tree species is known, then the weight of this volume can always be estimated.

For example, one approach tried in Kenya was to assess the standing volume by measuring individual trees in sample areas. Yield was estimated from notional rotations of the various tree species. These measurements excluded shrubs. It is possible to estimate increment and yield for individual species if growth rates, soil types and rainfall are known. This will require a detailed ecological survey.

The practical measurement of increment and yield of the total standing biomass is only possible by taking regular measurements (at least once a year, ideally from special inventory plots established at villages).

The next step is to collect information on wood consumption (see Chapter 5). This information is best obtained from surveys and questionnaires directed to the farmer and or the householder. The type of questions asked should include:

- how individual farmers manage their trees;
- the exact location of the trees, and the parts from which the biomass are collected;
- who has access rights to the biomass from these trees.

With the yield and consumption data collected, you can then create time-series data for production and consumption of biomass. Appendix 3.2 gives further insights into projecting supply and demand.

Hedges, shrubs and bushes

Hedges, shrubs and bushes are favourite collecting sites for woody biomass fuel, but standing measurements of such biomass is difficult. One solution is to measure height and crown and then cut and weigh a sample of shrub plants, together with the dead branches and twigs that collect underneath.

PROCESSED WOODY BIOMASS

Besides measuring the growing stock and annual yield of woody biomass and the annual production of plant and animal residues, it may be important to try to measure the supply of modified or processed biomass. Modification produces a waste product that can be burned, as in the dehusking of cereals and coffee, the conversion of sawlogs into sawn wood or the preparation of copra from coconuts. See Chapter 4 for further details.

Sawmill waste

Sawmill waste is a very important source of fuelwood as large quantities of waste can be generated. Sawmill waste can be divided into slabs or offcuts, bark (if separated from the slabs) and sawdust. The amount of waste produced from a sawlog depends on:

- the diameter of the log to be cut;
- the sawing method;
- the market for the sawn wood.

A frame saw is often slightly more efficient than a band saw, and both are more efficient than circular saws. Also, if there is a market for small pieces of sawn wood, this will decrease the amount of waste. Much depends also on the current market value of wood.

The quantity of biomass available after processing is therefore calculated from the input of raw material. The use to which it is put then depends on the market. For example, the slabs and offcuts may be consumed as a boiler fuel to drive the mill, or they may be used as the raw material for fibreboard or particle board production, making pellets, etc.

To determine uses, or potential uses, for this biomass it is necessary to carry out a sample survey at sawmills, board factories, pulp mills and other wood-consuming industries, classified by site, type and location. Sample surveys could be undertaken to determine the amount of such waste and its use.

Charcoal

Charcoal is produced by burning biomass, especially wood, in a restricted supply of air. It is an important urban household fuel in many developing countries. Charcoal is also used by industry for steel manufacture and blacksmithing. It is by far the most important processed biomass fuel. Charcoal production is a primary or secondary activity, for many millions of rural labourers in developing countries, and is one of the rural activities that brings in cash. Many charcoal makers are itinerants and often production may be semi-legal or illegal. It is important to record the legal status of charcoal making, as this may greatly hinder improved technology.

Charcoal is a major economic activity in many developing countries, particularly in Brazil, Africa and Thailand, which is expected to increase significantly in the future particularly in Africa, from about 22.5 Mtoe in 1995, 42 Mtoe in 2010 and about 58 Mtoe in 2020.[1] Even so, these estimates are very conservative; production estimates are very difficult to establish accurately because charcoal is an activity that is an integral part of the informal economy of many rural communities, in most cases.

Surveys should be undertaken to record the chain of activities and players from tree to market. This should include production sites, the technology employed, the people involved in the production, regulations, transportation and economics. Samples should be tested for carbon, moisture and ash contents, and energy value.

Various aspects are worth emphasizing with regard to traditional charcoal production, including the following:

- The enormous socio-economic importance of charcoal production and use in developing countries, for example, hundred of thousands, even millions, of people depend totally or partially on this activity. An estimated 200–300 million people use charcoal as their main source of energy around the world.

- The low energy efficiency and technology base (e.g. 12 per cent in Zambia, 11–19 per cent in Tanzania, 9–12 per cent in Kenya) results in considerable waste of resources as well as having serious environmental impacts. Recent data indicate that emissions are much higher than previously thought.[2]
- A major preoccupation in charcoal production is the slow pace of technological development. Indeed, this technology has remained, in the main, unchanged for centuries. One of the reasons is that charcoal is mostly an activity of the poor who are struggling to survive, let alone invest in technological improvements. There are very few countries that use charcoal for industrial purposes, such as Brazil, which might channel some resources into R&D, but even in these cases it is on a very limited scale.

Production technology, moisture content of the raw materials, chemical properties of the wood, and the skill of the operator all have a great influence on the quality and quantity of the charcoal. There is greater control over the process if charcoal is made in retorts or brick or steel kilns rather than earth kilns.

Apart from the moisture content of the wood, the type of wood and its chemical composition, the properties of charcoal depend significantly on the carbonization temperature. Carbonization at a low temperature produces a charcoal with a high level of volatiles known as 'soft and black' charcoal, which is mainly consumed in the domestic market. High-grade or 'white' is produced at very high temperatures and is used as a reducing agent in the iron-making industry.

It is impossible to obtain more than 50 per cent conversion by weight of wood to charcoal. In practice, the upper limit is about 30–35 per cent. If charcoal is carbonized properly, it consists of at least 75 per cent carbon by weight and has an energy value of approximately 32 MJ/kg, compared to around 20 MJ/kg for dry wood. Some 'charcoal' may turn out to be little more than charred wood with a carbon content just over 50 per cent. The energy value will then only be slightly more than dried wood. The charcoal yields from non-woody plants are slightly lower, as they only contain between 45 per cent and 47 per cent carbon.

The following steps should be followed in order to estimate charcoal production and the raw wood material required:

- Identify a good sample of charcoal producers.
- Record the type of kiln used by each producer.
- Estimate the quantity of wood utilized by each producer.
- Estimate the quantity of charcoal produced by each producer. This will depend on the moisture content and the density of the wood used to make it, the equipment used to make the charcoal and the skill of the producer. Note

that if yield is measured on a wet or air-dry (as opposed to oven-dry) basis (weight of charcoal output divided by wet weight of wood input), water increases the weight of wet wood in the denominator of the charcoal yield equation.

- Record the quantity of each type of charcoal produced – lump charcoal, powder and fines – and any other marketable products such as creosote.
- Record the uses to which the charcoal is put. The bulk of fines is usually left on site, although in some cases it is injected into a charcoal blast furnace or collected up and sold if there is a market. If charcoal is made from materials such as coffee husks, all the charcoal produced is either powder or fines.
- Record the production cycle, quantity of charcoal produced and time taken.
- Check wood consumption against growing stock and yield figures from the trees providing raw wood material.
- Calculate the yield obtained for each producer.

Separate columns for wood used as a raw material for charcoal production and the amount of charcoal produced should be used in the energy accounting system.

Knowing the input of the raw material and the output of saleable charcoal, a conversion factor can be determined. However, there is usually some waste produced between the producer and the marketplace, especially as it may be transported some distance. Some lump charcoal will powder and be lost or settle at the bottom of the bag and not be of much use for burning.

The bulk of charcoal is produced in earth kilns and is usually transported over long distances (100 km or more is common and in Brazil the distance can be as much as 1000 km). Thus, a conversion factor of 12 m^3 of wood raw material (about 8.5 t) per tonne of charcoal is more accurate than the FAO standard adopted by the UN of 6 m^3 per tonne of charcoal.

Charcoal is such an important fuel that it is essential that the conversion factors are as accurate as possible, especially when calculating woodfuel use in all its forms for energy. Conversion factors differ according to the type of technology and varying moisture contents (see Table 3.3). An important factor to be taken into consideration, for both fuelwood and charcoal, is distances over which they have to be transported. Transport can be a critical component in the accessibility factor.

To obtain high yields of good quality charcoal, the operator of an earth or pit kiln has to be skilful and alert. But despite the relative inefficiency of earth kilns, they are generally the most appropriate technology for the woodland, savannahs and rangeland areas where most of the charcoal is produced. Here the producer moves from site to site. Only when there is a concentrated amount of biomass within a small radius should a producer think of capital-intensive technology. Even with earth kilns, however, the efficiency can be improved by training and adopting certain techniques such as drying the raw material and constructing the kiln properly.

Table 3.3 *Conversion factors per tonne of charcoal sold*[a] *(average volume 1.4 cm/t at 15 per cent moisture content (db)*[b]

Kiln type	Unit m³ roundwood Moisture content: dry basis					
	15%	20%	40%	60%	80%	100%
Earth kiln	10	13	16	21	24	27
Portable steel kiln	6	7	9	13	15	16
Brick kiln	6	6	7	10	11	12
Retort	4.5	4.5	5	7	8	9

Notes:
[a] It is assumed that the fines are briquetted in the retort.
[b] With softwood about 60 per cent, more volume is required per tonne of charcoal and with dense hardwoods such as mangrove about 30 per cent less volume is required.

NOTES

1   The wood equivalent is approximately 70, 137 and 192 Mtoe (million tonnes oil equivalent), taking into account energy losses during the process of charcoal production. One toe corresponds, approximately, to 2.915 t of wood, but this can vary.

2   New studies in the past few years are indicating that the effects of charcoal production are much higher than predicted, for example approximately 0.65–1.41 CO_2 equivalent per kg of charcoal produced (see www.ecoharmony.com) which makes charcoal one of the most GHG-intensive. However, this is due to a combination of factors, one of which is the extreme inefficiency of charcoal production methods in many of the poorest countries.

REFERENCES AND FURTHER READING

Adlar, P. C. (1990) 'Procedures for monitoring tree growth and site change', *Tropical Forestry Papers*, no 23, Oxford Forestry Institute, Oxford, UK

Alder, D. (1980) *Forest Volume Estimation and Yield Prediction*, vol 2, FAO Forestry Paper no 22/2, FAO, Rome

Allen, A. B. (2004) *A Permanent Plot Method for Monitoring Changes in Indigenous Forests: A Field Manual*, Christchurch, New Zealand

Ashfaque, R. M. (2001) 'General position paper on national energy demand/supply', in *Woodfuel Production and Marketing in Pakistan*, Sindh, Pakistan, 20–22 October, RWEDP/FAO Report no 55, Bangkok, pp17–36

Bialy, J. (1979) *Measurement of Energy Released in the Combustion of Fuels*, School of Engineering Sciences, Edinburgh University, Edinburgh

Blyth, J., Evans, J., Mutch, W. E. S. and Sidwell, C. (1987). *Farm Woodland Management*, Farming Press Ltd, Ipswich, UK

Emrich, W. (1985) *Handbook of Charcoal Making*, D. Reidel Publishing Co., Holland

ETC Foundation (1990) *Biomass Assessment in Africa*, ETC (UK) and World Bank

Ford, E. D. and Newbould, P. J. (1970) 'Stand structure and dry weight production through the Sweet Chestnut (*Castanea sativa* Mill) coppice cycle', *Journal of Ecology*, 58, 275–296

Forestry Commission (1984) *Silviculture of Broadleaved Woodland*, Forestry Commission Bulletin 62, HMSO, London, UK

Forestry Commission (1988) *Farm Woodland Planning*, Forestry Commission Bulletin 80, HMSO, London, UK

Forestry Commission (1998) 'Sample plot summary data, computer printout. Mensuration Branch', Alice Holt Lodge, Surrey, UK (unpublished)

Hemstock, S. L. (1999) 'Multidimensional modelling of biomass energy flows', PhD thesis, University of London

Hemstock, S. L. and Hall, D. O. (1995) 'Biomass energy flows in Zimbabwe', *Biomass and Bioenergy*, 8, 151–173

Hollingdale, A. C., Krisnam, R. and Robinson, A. P. (1991) *Charcoal Production: A Handbook*, CSC, 91 ENP-27, Technical Paper 268, Natural Resources Council and Commonwealth Science Council (CSC), London

Husch B., Beers, T. W. and Kershaw, J. A. (2003) *Forest Mensuration, Measurement of Forest Resources Book*, John Wiley, 4th edn

IEA/OECD (1997) *Biomass Energy: Key Issues and Priorities Needs*, Conf. Proc. IEA/OECD, Paris

IEA/OECD (1998) *Biomass Energy: Data, Analysis and Trends*, Conf. Proc. IEA/OECD, Paris

Kartha, S., Leach, G. and Rjan, S. C. (2005) 'Advancing Bioenergy for Sustainable Development; Guidelines for Policymakers and Investors', Energy Sector Management Assistance Programme (ESMAP) Report 300/05, The World Bank, Washington, DC

King, G., Marcotte, M. and Tasissa, G. (1991) 'Woody Biomass Inventory and Strategic Planning Project (Draft Training Manual)', Poulintheriault Klockner Stadtler Hurter Ltd

Leach, G. and Gowen, M. (1987) *Household Energy Handbook: An Interim Guide and Reference Manual*, World Bank Technical Paper no 67, The World Bank, Washington, DC

Mitchell, C. P., Zsuffa, L., Anderson, S. and Stevens, D. J. (eds) (1990) 'Forestry, forest biomass and biomass conversion' (The IEA Bioenergy Agreement (1986–1989) Summary Reports), reprinted from *Biomass*, vol 22, no 1–4, Elsevier Applied Science, London

Openshaw, K. (1983) 'Measuring fuelwood and charcoal', in *Wood Fuel Surveys*, FAO 1983, pp173–178

Openshaw, K. (1990) *Energy and the Environment in Africa*. The World Bank, Washington DC

Openshaw, K. (1998) 'Estimating biomass supply: Focus on Africa', in *Proc. Biomass Energy: Data Analysis and Trends*, IEA/OECD, Paris, pp241–254

Openshaw, K. (2000) 'Wood energy education: An eclectic viewpoint', *Wood Energy News*, vol 16, no 1, 18–20

Openshaw, K. and Feinstain, C. (1989) *Fuelwood Stumpage: Considerations for Developing Country Energy Planning*, Industry and Energy Deptartment Working Paper – Series Paper no 16, The World Bank, Washington, DC

Philip, M. S. (1983) *Measuring Trees and Forests: A Textbook for Students in Africa, Vision of Forestry*, University of Dar es Salaam, Tanzania

Poole, A. (1998) Personal communication. Thoresby Estates Management Limited, Thoresby Park, Newark, Notts, UK

Prior, S. (1998) Personal communication, Oxford Forestry Institute, Oxford, UK

Ramana, V. P. and Bose, R. K. (1997) 'A framework for assessment of biomass energy resources and consumption in the rural areas of Asia', in Proc *Biomass Energy: Key Issues and Priorities Needs*, IEA/OECD, Paris, pp145–157

Rogner et al (2001) 'Energy resources', in *World Energy Assessment: Energy and the Challenge of Sustainability; Part II Energy Resources and Technology Options*, Chapter 5, UNDP, pp135–171

Rosenschein, A., Tietema, T. and Hall, D. O. (1999) 'Biomass measurement and moni-toring of trees and shrubs in semi-arid regions of Central Kenya', *Journal of Arid Environment*, 41, pp97–116

Rosillo-Calle, F. (2004) *Biomass Energy (Other than Wood)*, Chapter 10, World Energy Council, London, pp267–275

Tietema, T. (1993) 'Biomass determination of fuelwood trees and bushes of Botswana', *Forest Ecology and Management*, 60, pp257–269

Yamamoto, H. and Yamaji, K. (1997) 'Analysis of biomass resources with a global energy model' in *Proc. Biomass Energy: Key Issues and Priorities Needs*, IEA/OECD, Paris, pp295–312

APPENDIX 3.1 PROJECTING SUPPLY AND DEMAND

There are various methods for predicting supply and demand:

- constant-trend based projections;
- projections with adjusted demand;
- projections with increased supplies;
- projections including agricultural land, and
- projections including farm trees.

Constant-trend based projections

These assume that consumption and demand grow in line with population growth and that there is no increase in supplies. They provide a useful way to identify any resource problems and possible actions to bring supply and demand into a sustainable balance (see Table 3.4). Essentially, consumption grows with population at 3 per cent per year and supplies are obtained from the annual wood growth and clear felling of an initially fixed stock of trees. However, as wood resources decline, costs will increase and consumption will be reduced by fuel economics and substitution of other fuels.

Table 3.4 *Example of a constant trend-based projection: hypothetical wood balance*

	1980	1985	1990	1995	2000	2005
Standing stocks: m³[a]	7,500	16,010	13,837	10,827	6,794	1,520
Fuelwood yield: m³/yr[a]	350	320	278	217	136	30
Consumption: m³/yr[a]	600	696	806	935	1,084	1,256
Deficit: m³/yr[a]	250	376	529	718	948	1,226
Population[a]	(1,000)	(1,159)	(1,344)	(1,558)	(1,806)	(2,094)

Note: [a] 10×3.

Assumptions: Fuelwood yield: 2 per cent of standing stock (standing stock: 20 m³/ha).

Population: 1 million in 1980, growth at 3 per cent per year.

Consumption: 0.6 m³/capita/year.

Deficit is met by felling the standing stock.

Calculation method:

Calculations are performed for each year (t, t + 1, etc.), taking the stock at the start of the year and consumption and yield during the year:

Consumption (t) = Reduction in stock (t, t + 1) + Yield in year (t)

Stock (t) − Stock (t + 1) + M/2 × [Stock (t) + Stock (t + 1)]

where M = Yield/Stock expressed as a fraction (0.02 in this case).

Hence, to calculate the stock in each year: Stock (t + 1) × (1 − M/2) = Stock (t) × (1 + M/2) − Consumption (t).

Projections with adjusted demand

This is a useful step for examining reductions in per capita demand and its effects on declining wood resources. The adjustments can then be related to policy targets, such as improved stove programmes or fuel substitution.

Projections with increased supplies

Wood supplies can be increased by a variety of measures, for example, better management of forests, better use of wastes, planting, use of alternative sources such as agricultural residues, etc. Targets for these additional supply options can easily be set by estimating the gap between projected woodfuel demand and supplies.

Projections including agricultural land

In most developing countries the spread of arable and grazing land, together with commercial logging in some areas, is a major cause of tree loss. When land is cleared by felling and burning (in situ), then this results in greater pressure on existing forest stocks for fuelwood. If the wood cleared is used for fuel this will contribute to relieving this pressure.

Projections including farm trees

Trees have multiple uses (e.g. fruit, forage, timber, shelter, fuelwood, etc.). These farm trees, which are fully accessible to the local consumers, are often a major source of fuel in many rural areas and hence should be included in any projection models.

Further reading

Leach, G. and Gowen, M. (1987) *Household Energy Handbook: An Interim Guide and Reference Manual*, World Bank Technical Paper No. 67, World Bank, Washington, DC, pp132–140

Openshaw, K. (1998) 'Estimating biomass supply: Focus on Africa', in *Proc. Biomass Energy: Data Analysis and Trends*, IEA/OECD, Paris

APPENDIX 3.2 MEASURING FUELWOOD
RESOURCES AND SUPPLIES

In estimating total wood resources and actual or potential wood supplies it is first necessary to make a clear distinction between standing stocks and *resource flows*, that is, the rate of wood growth or yield. Other important distinctions for energy assessment include: competing uses for the wood (e.g. timber, construction poles, etc.); the fraction of the standing stock and yield that is actually accessible for exploitation due to physical, economic or environmental reasons; the fraction of the yield that can be cut on a sustainable basis; and the fraction of the cut wood that is actually recovered.

Estimating stock inventories

The standing stock of trees is normally estimated by aerial surveys or satellite remote sensing to establish the areas of tree cover by categories, such as closed forest, plantations, etc. This information is then combined with the survey and mensuration data.

It is important to be aware that estimates of tree stocks are always approximate. Most inventory data are for the commercial timber volumes which are a small proportion of total standing biomass. The quantity and also the quality of fuelwood biomass may greatly exceed the commercial timber volume.

Estimating supplies

Table 3.5 shows the estimated amount of wood that can be obtained from natural forest:

- by depleting the stock, and
- by sustainable harvesting.

Essentially, the method involves simple multiplication to adjust stock and yield quantities by the accessibility and loss factors. This model could apply equally well to a managed plantation or village woodlot (although, of course, with different numbers), to estimating the effects of forest clearance for agriculture (partial or complete stock loss), or to evaluating the impact of fuel gathering on forest stocks.

Table 3.5 *Example of a stock and yield estimation method: natural forest/plantation (hypothetical data)*

Assumptions	Stock data	Yield data
Supply factors		
Forest area		1000 ha
Stock density		200 m³/ha
Stock volume		200,000 m³
Mean increment		0.4 m³/ha/yr
Sustainable yield		3.8 m³/ha/yr
Gross sustainable yield (A × F)		3,800 m³/yr
Fraction available for fuelwood	0.4	0.4
Fraction accessible	0.9	0.9
Harvest/cutting fraction	0.9	0.9
Gross sustainable harvest (G × I × J)		3,078 m³/yr
Fuelwood sustainable harvest (K × H)		1,231 m³/yr
Clear felling		
Gross harvest (C × I × J)		162,000 m³
Fuelwood harvest (M × H)		64,800 m³
Wet density (0.8 t/m³)		
Net heating value (15 GJ/t or MJ/kg)		
Energy harvest: clear felling (N × O × P)		777 TJ[a]
Energy harvest: sustainable (L × O × P)		14.6 TJ/yr[a]
		14.6 GJ/ha/yr
Other wood: clear felling (M – N) × O		77,700 t
Other wood: sustainable harvest (K – L) × O		1477 t/yr
		1.47 t/ha/yr

Note: [a] TJ = terajoule = 1000 GJ.

Further reading

Leach, G. and Gowen, M. (1987) *Household Energy Handbook: An Interim Guide and Reference Manual*, World Bank Technical Paper no 67, The World Bank, Washington, DC, pp93–94

Ramana, V. P. and Bose, R. K. (1997) 'A framework for assessment of biomass energy resources and consumption in the rural areas of Asia', in Proc. *Biomass Energy: Key Issues and Priorities Needs*, IEA/OECD, Paris, pp145–157

Yamamoto, H. and Yamaji, K. (1997) 'Analysis of biomass resources with a global energy model' in *Proc. Biomass Energy: Key Issues and Priorities Needs*, IEA/OECD, Paris, pp295–312

APPENDIX 3.3 TECHNIQUES FOR MEASUREMENT OF TREE VOLUME

The most common diameter measurement taken in forestry is the main stem of standing trees. This is important because it is one of the directly measurable dimensions from which tree cross-sectional area and volume can be computed. On standing trees the most common diameter measured is the reference diameter usually measured at 1.3 m above the ground level and commonly known as the diameter at breast height (DBH).

There are various instruments for measuring DBH. The most common include callipers, diameter tape, the Wheeler pentaprism and the Bitterlich relascope (or spiegel lelaskop). The first two are cheap and widely available and the last two are more specialized to foresters and available from specialist suppliers.

Bark thickness

Whether inside bark (ib) or outside bark (ob) diameter measurements are taken depends on the purpose for which the measurements are made. The proportion of bark volume compared to the total volume with bark varies from a few per cent to about 20 per cent for the majority of species. Bark thickness on standing trees can be determined with a bark gauge.

Calculation of volumes

Using measured volumes
The volume of each tree tallied in the inventory is calculated by substituting the recorded DBH (ob) and stem bole length into the formula:

$$V = aD^bH^c$$

These volumes are tabulated by species and by diameter class, totalled and averaged for each stratum. These average volumes by species and diameter class are used in calculation of the Stock Table for each stratum.

Two-entry volume table
If desired, it is possible to develop a two-entry volume table, i.e. DBH and stem bole length values (probably 0.5 m) into the derived formula. The results are tabulated to produce a 'look up' table showing volumes by diameter class and height class (see Table 3.6).

Average volume tables
Another method of determining average volumes by diameter class is by calculating for some species the average diameter and average height by diameter class; plotting this data as a graph and fitting a least squares curve: substituting the height and diameter value from this curve into the calculated formula $V = aD^bH^c$ to obtain a table showing:

DBH class   Volume
(cm)   (m)

Compilation of branch volume
- Step 1: calculate the volume of each branch section using the formula:

$$Volume = \frac{(A_1 + A_2)L}{2}$$

where:

A_1 = cross-sectional area (ob) of the bottom end of the branch section.
A_2 = cross-sectional area (ob) of the top end of the branch section.
L = length of the branch section.

- Step 2: sum the volumes per tree.
- Step 3: list the trees by species, DBH class and branch volume.

Table 3.6 *An example of a two-entry volume table*

Diameter class (cm)	Height classification (m)					
	1.0	1.5	2.0	2.5	3.0	etc.
	Volume (m³)					
20						
25						

- Step 4: perform a multiple regression analysis in order to find the best fit equation for the relationship between DBH class and branch volume.
- Step 5: tabulate the equation results.
- Step 6: prepare a branch volume summary for each stratum by multiplying the number of trees per hectare (from the Stand Table) by the branch volume (from the table prepared in Step 5) for each DBH class.

Compilation of tree volume
This process consists of calculating the volume of the tree stem bole from the individual section measurements taken during the volume and Defect Study. Individual tree bole volumes are used in the compilation of the sample plot volumes, calculated on a species basis.

Volume of felled trees
- Step 1: calculate the gross volume of each section of the tree stem bole using the simple formula:

$$Volume = \frac{(A^1 + A^2)L}{2}$$

where:

A^1 = cross-sectional area (ib) of the bottom of the section.
A^2 = cross-sectional area (ib) of the top end of the section.
L = length of the section.

- Step 2: sum the section volumes per tree.
- Step 3: list tree data by species showing diameter, height and volume.

Further reading

Ashfaque, R. M. (2001) 'General position paper on national energy demand/supply', in *Woodfuel Production and Marketing in Pakistan*, Sindh, Pakistan, 20–22 October, RWEDP/FAO Report no 55, Bangkok, pp17–36

King, G., Marcotte, M. and Tasissa, G. (1991) 'Woody Biomass Inventory and Strategic Planning Project' (Draft Training Manual), Poulintheriault Klockner Stadtler Hurter Ltd, pp11.1ff

Openshaw, K. (1998) 'Estimating biomass supply: Focus on Africa', in *Proc. Biomass Energy: Data Analysis and Trends*, IEA/OECD, Paris, pp241–254

Appendix 3.4 Measuring fuelwood and charcoal

Measuring fuelwood

Much fuelwood is collected by the headload. A number of headloads could be measured either by weight or volume and an average for a particular district or country should also be established. The headload size may differ considerably from district to district, region to region, etc. For example, in Tanzania the average of headload was about 26 kg whereas in Sri Lanka it was about 20 kg. Therefore, size and weight must be established for each particular case.

Volume and weight

These two methods of measuring each have drawbacks. If volume is used, then the conversion factor from the bundle to solid measure can vary enormously, depending on whether the headload consists of one large log or many small branches.

In some countries the stere, or stacked cubic metre, is the standard measure, but in using this measure it is not possible to know the correct conversion factor to apply. If the stere is made up from bundles, then the conversion factor will be much lower than if it is made up from stacked stemwood; although the stacked measure is not an exact measure and can be up to 20 per cent more than a true stere. This also applies to the other stacked cubit feet or the metric core (see Appendix IV). One advantage of the volume measure over the weight measure is that the volume of wet wood does not differ greatly from air-dry wood (up to approximately 5 per cent difference). If a standard conversion factor to convert weight into volume is used, without accounting for the moisture content of the wood, there can be a 100 per cent difference in volume estimation, depending on whether the wood is green or oven dry.

It is important to note that in the case of domestic fuelwood, the volume is not such a suitable measure to quantify. The wood is often of an irregular shape, and as the quantities used are relatively small, it is usually much easier to determine the weight. Weight may therefore be a much more convenient measure to use to ascertain solid volume, for the weight of a bundle of wood is easier and quicker to determine than trying to determine the gross volume of an irregular shaped headload of fuelwood. If the solid volume were to be measured, then every piece of wood would have to be measured separately, or the water displacement method used. It is, however, important to know the moisture content of wood if weight is the measure for assessing the solid roundwood volume. An additional problem with using weight as a measure to determine solid volume is that the weight depends on density, and the density within and between wood species is not uniform. Only if the moisture content is the same, is the energy given off from a piece of wood more or less the same on a weight basis, irrespective of species. In other words, the energy content per unit weight for different species of wood varies far less than the energy content per unit volume.

Measuring charcoal

In the production of charcoal care must be taken to produce the quality of charcoal required. A few methods have been developed to analyse the raw materials and products of the charcoal-making process (e.g. sample preparation techniques, testing of physical properties and chemical analysis), details of which can be found in Hollingdale et al, 1991. Chemical analysis is particularly important. For example, to find the Gross Calorific Value (GCV) of a charcoal, a known quantity is burned under strictly controlled conditions in oxygen to ensure complete conversion of the charcoal to its combustion products. The heat released by this combustion is determined on the basis of the following equation:

$$\text{Heat release} = \text{mass of apparatus} \times \text{specific capacity of the apparatus} \times \text{temperature rise}$$

Equally, the moisture content of a charcoal sample (which represents the water that is physically bound in it) can be found by driving off free moisture from a sample in an oven and recording the mass loss.

Charcoal is usually sold by volume – per standard bag or basket, per tin or per pile – but sometimes directly by weight. Most frequently charcoal is sold by the standard bag, which can vary from area to area and country to country. The weight of charcoal depends on the moisture content and on the density of the parent wood, assuming it has been completely or near completely carbonized.

It is important therefore to know the species from which charcoal is made. For example, charcoal made from tropical hardwoods (with a volume of about 1.4 m^3 and 15 per cent mc) will weigh approximately 33 kg per bag, whereas a bag of charcoal made from softwood will weigh, on average, about 23 kg.

When charcoal is sold by the tin, this can vary in size. For example, a 20 litre paraffin tin will contain about 7 kg of charcoal if tropical hardwood is used. The tin and the bag are sold at prices that fluctuate according to season, inflation, etc., while a pile is usually sold at a fixed price and therefore the quantity of the pile varies from season to season and over time.

Charcoal efficiency can be defined either in terms of weight or energy.

$$\text{weight} = \frac{\text{charcoal output (kg)}}{\text{wood input (kg)}}$$

$$\text{energy} = \frac{\text{charcoal output (MJ)}}{\text{wood input (MJ)}}$$

It is important to note that the heating value of the primary end product (charcoal in this case) is determined by its carbon content. The formula for the relationship between carbon content (C) and higher heating value (HHV) (dry basis) of combustible fuels can be described as follows:

$$HHV = 0.437 \times C - 0.306 \text{ MJ/kg}$$

Converting charcoal to roundwood equivalent

Three major problems arise if charcoal is converted back to roundwood equivalent: wood density, moisture content of the wood and conversion method, all of which need to be known before conversion can be carried out. The density of the wood governs the yield of charcoal and thus a given volume of charcoal will give different weights of charcoal, depending on the species, moisture content, technology, etc. Moisture content also has an important effect on the yield of charcoal; as note above, the drier the wood the greater the yield of charcoal. The method used to produce charcoal can also affect the yield considerably. For example, the range for average tropical hardwoods at 15 per cent moisture content can be from about 4.5 m^3 per tonne in a metal retort to about 27 m^3 per tonne at a 100 per cent moisture content in a poorly designed kiln. Most of the charcoal produced in the majority developing countries is produced in earth kilns whose conversion factor can vary from about 10 m^3 per tonne of charcoal up to 27 m^3 (see Openshaw, 1983).

Further reading

Bialy, J. (1986) *A New Approach to Domestic Fuelwood Conservation: Guidelines for Research*, FAO, Rome

Emrich, W. (1985) *Handbook of Charcoal Making*, D. Reidel Publishing Co., Holland

Hollingdale, A. C., Krisnam, R. and Robinson, A. P. (1991) *Charcoal Production: A Handbook*, CSC, 91 ENP-27, Technical Paper 268, Natural Resources Council and Commonwealth Science Council (CSC), London, pp93 ff

Openshaw, K. (1983) 'Measuring fuelwood and charcoal', in *Wood Fuel Surveys*, FAO 1983, pp173–178

4

Non-woody Biomass and Secondary Fuels

Frank Rosillo-Calle, Peter de Groot, Sarah L. Hemstock and
Jeremy Woods

INTRODUCTION

Non-woody biomass includes:

- agricultural crops;
- crop residues;
- herbaceous crops;
- processing residues;
- animal wastes;

and also:

- densified biomass (briquettes, pellets, wood chips), which are increasingly being traded internationally;
- secondary fuels (biodiesel, biogas, ethanol, methanol and hydrogen);
- tertiary fuels, as their development can have major impacts on biomass resources.

This chapter also assesses animal traction, which still plays a major role in many countries around the world and also impacts on biomass resources.

There is no clear-cut division between woody and non-woody biomass. For example, cassava and cotton stems are wood, but as they are strictly agricultural crops it is easier to treat them as non-woody plants. Bananas and plantains are often said to grow on 'banana trees', although they are also considered as agricultural crops. Coffee husks are treated as non-woody residues, whereas the coffee clippings and stems are classed as wood. Classification into woody and non-woody biomass is for convenience only, and should not dictate which data are collected. It is only important to gather data about agricultural crops and residues that are used for fuel, not the total non-woody biomass production on any given site. Many plants are unsuitable as fuel and most have alternative uses.

The methodology employed for estimating non-woody biomass depends on the type of material and the quality of statistical data available or deducible. Where fine detail is required for the implementation of projects, exhaustive investigation is necessary to provide a clear, unambiguous report on each type of biomass resource, and to provide a detailed analysis of the availability, accessibility, ease of collection, convertibility, present use pattern and future trends.

In the case of crop and processing residues, residue indices must also be determined in addition to the above parameters, and the alternative uses of such residues must be analysed.

Agricultural and agro-forestry residues as fuel

Most farming systems produce large amounts of residues that offer a large potential for energy that is currently greatly underutilized in many parts of the world. It is only in wood-scarce areas that raw agricultural residues are often the major cooking fuels for rural households. The greatest concentration of residue-burning has been in the densely populated plains of Northern India, China, Pakistan and Bangladesh, where as much as 80–90 per cent of household energy in many villages comes from agricultural residues.

The use of crop residues is changing rapidly. For example, in many areas of China, rapid economic growth has led to the swift replacement of traditional biomass due to fuel switching from crop residues to coal and other fossil fuels. This is causing environmental problems, such as fire hazards when, for example, residues are left to rot in the fields or simply burned in situ. In other countries, particularly industrial countries, the energy use of such residues is increasing for modern applications. For example, in the UK almost all straw residue surpluses are now used in combustion plants to generate heat and electricity.

Determination of yield from crop residues

Accurate estimates of the availability of crop residues require good data on crop production by region or district. If these data are not available, a survey will be necessary. A survey should include information on all the uses for crop residues besides fuel (burning in situ, mulching, animal feed, house building, etc.) so that the amount available as fuel can be calculated.

Crop residues are usually derived from parts of the plant growing above ground. Exceptions include groundnuts and sometimes part of cotton crop residues. Some communities may also burn roots.

Table 4.1 gives examples of various types of crop residues at various sites.

An assessment of agricultural residues should include the following steps.

Define what is meant by non-woody biomass
As already indicated in the Introduction, there is no clear-cut division between

Table 4.1 *Various types of crop residues from different sites*

Crop	Field (standing)	Field (cut)	House	Factory
Subsistence/cash				
Cereals				
Maize	Stover and leaf	Cob leaves	Cob	Parchment
Deep water paddy	Straw (nara)	Straw (kher)	Kher	Husk
Normal rice paddy	Stubble	Straw	Straw	Husk
Millet; sorghum	Straw	–	Chaff	–
Wheat, etc.	Stubble	Straw	–	–
Cassava	–	Stem	–	Waste
Pulses	Stem	–	–	–
Plantain, banana	–	Stem	Fruit stem	–
Papyrus	Stem	–	–	–
Heather, etc.	Whole plant[a]			
Cash crops				
Coffee (dry process)	(Woody biomass)		Cherries[b]	Cherry, husk
Coffee (wet process)	(Woody biomass)		–	Cherry, husk
Cotton	–	Roots and stems[c]	–	(Tow)
Coconut; palm nut	(Wood)	Fronds	Husk and shell	Husk and shell
Nut trees	(Woody biomass)		Shell	Shell
Groundnut	–	Stems		Shell
Sugar cane	–	–	–	Bagasse
Sisal	–	Old plants	–	Waste
Jute; kenaf; flax	–	Waste	–	Waste
Pineapple	–	Old plants	–	Waste
Indirect use				
Grasses[d]	(Grass)	(Hay)	–	–

Notes:
[a] In some countries heather-type plants are uprooted from upland areas, dried and burned by householders.
[b] Coffee cherries make a good fertilizer.
[c] Cotton stems and roots have to be uprooted and removed or destroyed within two months of harvest because of pathogens and nematode problems.
[d] Grasses have recently been used as fuelwood either mixed with straw or just on its own. More recently grasses such as Miscanthus, are being considered as potential energy plantations.

woody and non-woody biomass. Classification into woody and non-woody biomass is for convenience only and should not dictate which data are collected.

It should also be borne in mind that only those agricultural crops and residues that are used for fuel should be considered rather than the total non-woody biomass production on any given site as many plants are unsuitable for use as fuel and most have alternative sites.

Obtain data on yields and stocks

For agricultural crops, dependable information on yields and stocks, quantified accessibility, calorific values, and storage and/or conversion efficiencies must be accurately determined. A study of the socio-cultural behaviours of the inhabitants of the project area will help to determine use patterns and future trends.

Note that for crop residues there is usually no stock, and the yield is the amount generated per annum.

Calculate potential quantities of residues

Agricultural crops are grown either commercially or for subsistence. It is likely that little or no information will be available for subsistence crops, so it will be necessary to collect data, possibly using remote sensing techniques. Total production can then be calculated using existing data on the yields of the various crops (though this data is also often of poor quality). See Appendix 2.1 Residue calculations for further details on how to calculate residues.

The amount of above-ground biomass produced by crops is usually one to three times the weight of the actual crop itself. Estimates have been made of these residues in various countries. Unfortunately, the potential of these residues has not been systematically inventoried. As a result, the quantity of residues has been calculated via estimates of the ratio of by-product to main crop yields for each crop type and the relation between crop and by-product, and by multiplying the crop production of a particular year by the residue ratio, i.e. in the case of wheat 1.3 times as much wheat straw is produced compared to the grain yield, depending on the variety.

Another method for estimating crop residues is to use the crop residue index (CRI). This is defined as the ratio of the dry weight of the residue produced to the total primary crop produced for a particular species or cultivar. The CRI is determined in the field for each crop and crop variety, and for each agro-ecological region under consideration. It is very important to state clearly whether the crop is in the processed or unprocessed state. For example, in the case of rice, is the husk included in the crop weight? If the residue has other uses besides energy, a reduction factor should be applied.

The quantity of biomass from crop residues that is available for fuel is only a fraction of this total, as not all of it will be accessible.

To obtain accurate estimates of residues production it is therefore important to have good estimates of crop production by country, region or district. This may entail undertaking surveys, especially in the subsistence sector, to determine production of both crops and plant residues, and should include all possible uses of residues in addition to fuel. If only general estimates of crops residues are required, crop production figures may be obtained from country statistics or UN bodies such as FAOSTAT. However, such statistics maybe based on guesses when dealing with subsistence agricultural production and hence, if accurate information is needed, field surveys may be necessary.[1]

Identify alternative uses
There are different types of residues (agriculture, forestry, animal, etc.) with very different characteristics and potential end-use applications. Many of these residues are usually underutilized and, in theory, there are considerable opportunities to use them as an energy source. In practice, however, this potential is often not realized, not only due to availability but also due to other factors such level of socio-economic development, cultural practices, etc.

Despite their potential as an energy source, agricultural residues have to compete with other alternative uses, particularly with the need to preserve soil fertility, retain moisture and provide soil nutrients, as well as various other uses of which fodder, fibre and fuel are the most common. Socio-economic changes also mean that consumer preferences change and while in some areas (e.g. China) these residues are of less value as a source of energy, in other countries such as the UK they are used to provide modern services.

Record the site of production and use
When measuring crop residues, it is advisable to record the site at which it is produced (in the field, where it may be left standing or cut, in the house or in the factory). This information is important as the further away from the consumption centre, the less likely the residue is to be used.

Calculate the accessibility of the residue
The quantity of residues actually available for use is only a fraction of the total due to issues of accessibility. Accessibility of crop residues depends mainly on the location and the economic value of the residue. The location determines the collection and transport costs. If this cost is greater than the economic value of the residue, it will not be used for fuel.

PROCESSED RESIDUES

Processed residues such as bagasse and rice husks are often important sources of biomass fuels. First, identify the relevant industries and the types of waste they produce. Then collect information on the type, composition (solid or liquid), quantity and dispersal (from intensive or extensive sources) of the waste. Statistical data should be available from the industry concerned, for example, from waste collection firms. For important sources of wastes this data can be verified in the field at a later date. Of greatest interest are intensive industries producing the larger quantities of waste. These sources should be investigated first.

It should be borne in mind that the use of biomass is continuously changing and what in the past was of little value can become economically important, and vice versa. This is illustrated by the increased use of bagasse and poultry litter for fuel.

Bagasse

About 350–600 Mt of sugar cane bagasse are produced annually worldwide, most of which was wasted in the past. However, in recent years bagasse has become increasingly valuable as a source of energy, thanks to new technological developments (such as improved integrated biomass gasifier/gas turbine (IBGT) systems for power generation, gas turbine/steam turbine combined cycle (GTCC) systems and increased political support. Many sugar-producing countries have (or are planning) co-generation programmes (e.g. Brazil, India, Mauritius). Thus, the use of bagasse has been transformed from an undervalued residue that needed to be burned inefficiently just to get rid of it, into a source of considerable economic value.

For example, a study by Larson and Kartha (2000) has shown that in developing countries as a whole, 'excess' electricity (i.e. above and beyond that needed to run the sugar/ethanol mill), could amount to 15–20 per cent of the projected electricity generation in 2025, or about 1200 TWh/yr, out of a total production of over 7100 TWh. Moreira (2002) has also identified a potential of 10,000 TWh/yr from a potential area of 143 Mha of sugar cane around the world.[2] Appendix 4.1 discusses how Mauritius set up its cogeneration programme using sugarcane bagasse.

Poultry litter

Another good example is the use of poultry litter in combustion plants in the UK. Poultry litter is the material from broiler houses and contains material such as wood shavings, shredded paper or straw, mixed with droppings. As received, the material has a calorific value of between 9–15 GJ/t, with variable moisture content of between 20 and 50 per cent depending on husbandry practices. There is about 150 MW installed capacity worldwide (75 MW in the UK and over 50 MW in the United States) and this is growing rapidly. This represents a new economic, energy and environmental benefit of a resource that in the past was mostly wasted, brought about mostly by environmental pressures (Rosillo-Calle, 2006).

When dealing specifically with processed residues it is important to prepare flow diagrams showing:

- the point of origin of the waste;
- the quantity of waste available from each site throughout the year, with historical data if possible;
- the composition of the waste;
- the methods by which the waste is collected;
- the means by which the waste is transported;
- the destination and use (if any) or disposal of the waste.

Such a flow chart will allow the most valuable and available wastes to be listed in order of their priority as sources of energy (see Chapter 2 for more information on flow charts).

ANIMAL WASTES

The use of animal residues, particularly dung, is declining except in the case of some larger farms that use it to produce biogas, etc. The potential for energy from dung alone has been estimated at about 20 EJ worldwide (Woods and Hall, 1994). However, the variations are so large that figures are often meaningless. These variations can be attributed to a lack of a common methodology, which is the consequence of variations in livestock type, location, feeding conditions, etc. Nonetheless, there are some general rules that can be applied to give overall estimates.

However, it is increasingly questionable whether animal manure should be used as an energy source on a large scale, except in specific circumstances, for example:

- manure may have greater potential value for non-energy purposes (i.e. if used as a fertilizer it may bring greater benefits to the farmer);
- dung is a poor fuel and people tend to shift to other, better quality biofuels whenever possible;
- the use of manure may be more acceptable when there are other environmental benefits (i.e. the production of biogas and fertilizer, because there are large surpluses of manure which, if applied in large quantities to the soil represent a danger for agriculture and the environment, as is the case in Denmark);
- environmental and health hazards are much higher than other biofuels (Rosillo-Calle, 2006). However, in some cases animal residues are increasingly being used mixed with straw and other farm wastes in industrial applications.

Are animal wastes worth assessing?

Measuring supply and consumption will not be useful unless dung is an important source of energy. Enquiries, observations and demand surveys will indicate the importance of dung as a fuel, but this process takes time and depends on variables that can be hard to assess. Only proceed if animal wastes are a source (or potential source) of energy. Dung is not an important fuel in all countries, although in other countries, for example, the Indian sub-continent and Lesotho, dung is one of the primary fuels.

If your initial assessment suggests that animal wastes should be surveyed, you should carry out the following steps.

Determine the number of animals
Determine the number of animals from which the dung is obtained. Reliable estimates of animal numbers are often not available. Be aware of the many pitfalls in calculating animal waste for reasons indicated above.

Calculate the quantity of dung produced
An assessment of animal wastes will involve:

1   a census of animal numbers by species, and by region or district, and
2   an estimate of their average weight.

More refined data will include information on age and gender. Then use 1 and 2 above to obtain estimates of the average quantities of dung produced, preferably over at least one year.

The dung produced by fully-grown animals is proportional to the quantity of food they eat. Food intake is roughly related to the size and weight of the animal, which will vary according to the particular country or region. However, type and quality of feed are also significant. During the dry season both the quantity and the quality of the feed may decrease, resulting in reduction in the wastes produced. Thus, both seasonal and regional feed variation must be taken into account when calculating average or total droppings. Weight and moisture content are the most important data to record at each site, particularly the moisture content at the time of burning and whether the dung is mixed with any other biomass such as straw, as this will affect the energy value.

If the region is prone to droughts, or the climate is erratic, estimates over a longer period should be made. The total animal waste available is the product of the total produced (P), taking into account accessibility factors (A), the collection efficiency (E) and the feed variation factor (F).

In practice, it is not usually necessary to calculate production per 500 kg of live weight. The amount of waste produced can be calculated directly from a census and from information in the literature. Key data can be verified in a field survey. The weight and moisture content of dung should ideally be measured at each production site.

Calculate accessibility

The accessibility factor for animal dung may vary from zero to one. For housed animals, such as pigs in piggeries, and intensively farmed animals, it can be assumed that waste is 100 per cent collectible and accessible. A census should be carried out for all intensively reared and housed animals. The total droppings for these animals can be calculated and this figure can be taken as representing the potential dung supply.

For extensively farmed animals (the great majority), estimation of accessibility and ease of collection are more difficult. Methods are similar to those proposed for crop residues. The collection efficiency takes into account those animal droppings which, even though they are accessible, cannot all be collected – which may be a large proportion. The collection efficiency is obtained by a survey to estimate the ratio of the amount collected to the total estimated droppings.

Ascertain other uses for the wastes

Dung has many other uses, such as in house building as a binding agent or to coat walls and floors. Some farmers consider dung so valuable as a fertilizer as indicated, that they do not use it for any other purpose. However, dung largely loses its value as a fertiliser after it is left in the sun to dry for a few days, although it may still be useful as a soil conditioner. These factors make it difficult to assess both the production of dung and its availability as fuel.

As with industrial wastes, it is useful to develop a flow diagram incorporating the following information:

- the size and distribution of animal-rearing enterprises;
- the yields of waste over all seasons, with historical data if possible;
- the moisture content when fresh and when collected;
- the present means of collection, uses (particularly on the land) and disposal.

Use the above to calculate the availability for bioenergy.

Tertiary wastes

These wastes are becoming increasingly by attractive, not only because of the large quantities available, but also for environmental reasons. Enormous amounts of municipal solid wastes (MSW) are generated annually that were largely ignored in the past. Rognar et al (2001) estimated the global economic energy potential from MSW at about 6 EJ (c.138 Mtoe). However, in the past few years important changes have taken place that make MSW more attractive. MSW is now even recognized by some authorities as a renewable energy (RE) resource, although this is still debatable. It is rather complicated to provide reliable figures on the amount of MSW generated because of the differences among countries, and among rural and urban dwellers within countries (e.g. 314 kg/yr per capita in Japan, 252 kg/yr in Singapore or 170 kg/yr in Brazil). However, there is no doubt that MSW is a sizeable resource that can be converted by incineration, gasification or biodegradation to electricity, heat and gaseous and liquid fuels, and thus should be taken into consideration (see Rosillo-Calle, 2004).

HERBACEOUS CROPS

The use of herbaceous crops as an energy source is not new, and in some areas of China it has been normal practice. These traditional practices, however, are gradually being phased out as better alternatives become available. However, some new species are now being proposed for commercial energy production. These species are more attractive to farmers because they can be harvested annually using the same machinery as food crops. There are various herbaceous crops being considered as suitable candidates for energy. For example, miscanthus (*Miscanthus × giganteus* and *Miscanthus sacchariflorus*) reed canary grass (*Phalaris arundinacea*) and switchgrass (*Panicum virgatum* L) have extensively been evaluated as potential energy sources in the United States and some EU member countries. It should be borne in mind that such crops are site specific and thus much will depend on local conditions. These crops are briefly described below.

Miscanthus

Miscanthus has been one of the most extensively studied crops and therefore considerable knowledge has been gained that has taken this crop from an experimental concept to the verge of commercial exploitation. As an energy crop, miscanthus differs from short rotation coppice (SRC) in that it can be harvested annually. A further advantage is that all aspects of its propagation, maintenance and harvest can be done with existing agricultural machinery. Long-term yield averages at the most productive sites have been in excess of 18 t/ha/yr, with few agrochemical inputs. Most economic analyses so far indicate that the only viable market for miscanthus feedstock is electricity generation, and in particular in co-firing with coal power plants (Bullard and Metcalfe, 2001).

Reed canary grass

Reed canary grass is a C3 seed-producing species with similar characteristics to switchgrass, and is well adapted to cool and wet conditions. Yields are comparable to SRC, with energy characteristics similar to straw, and thus it can be used as fuel in modern combustion plants. However, only limited research has been done so far and its possible fuel use should only be considered in the long term. The main advantages of reed canary grass are its high potential yields, similar caloric values per unit weight to wood and its non-requirement of any specialized machinery (see Bullard and Metcalfe, 2001).

Switchgrass

The potential of switchgrass (a C4 grass) as an energy source has been extensively researched in the United States, where it is very well adapted to most Northern America conditions. This grass can be directly burned or mixed with other fuel sources such as coal, wood, etc., and is therefore very suitable for co-firing. Switchgrass grows in a thick cover and can reach about 2.5 m high and it is not only a good energy source but also provides an excellent cover for wildlife and prevents soil erosion. It is a long-lived perennial plant that can also produce high yields on marginal soil with a low establishment cost. It is cold tolerant, requires low inputs and is adaptable to wide geographical areas. However, much still needs to be learned at a practical, site-specific level to be able to consider this grass as a significant possible source of energy.

What is important is that the use of herbaceous crops in the future will be in modern applications and thus, unlike traditional biomass energy applications, commercial measuring techniques (mostly from agriculture) will apply.

SECONDARY FUELS (LIQUID AND GASEOUS)

Secondary fuels obtained from raw biomass are increasingly being used in modern industrial applications, particularly transport, and therefore they will play a greater role in providing energy in the future. In fact some of them will play a key role and thus are discussed briefly in this chapter. However, it is important to bear in mind that most secondary fuels are used in modern applications and thus commercial measurement techniques apply.

Biodiesel

The use of biodiesel has increased significantly in the past decade, particularly in the EU, which is leading the world in this field with an estimated production of about 1.5 Mt in 2004 (Germany alone represents about 1.2 Mt) (see also the case study on biodiesel in Chapter 7). Growth is expected to continue in many other countries with some potential major programmes in the pipeline (Brazil, EU-wide, the United States, Malaysia, etc.).

Considerable advances have already been made in biodiesel production and use including:

- diversification of feedstock;
- process technology and fuel standards ensuring higher fuel quality;
- better marketing;
- diesel engine warranties;
- legal measures support.

A study by the Austrian Biodiesel Institute (ABI), www.abi.at, identified rape-seed as the most important source with a share of over 80 per cent, followed by sunflower oil with over 10 per cent, mostly used in Italy and southern France. Soybean oil is preferred in the United States. Other raw materials are palm oil in Malaysia, linseed and olive oil in Spain, cotton seed oil in Greece, beef tallow in Ireland, lard and used frying oil (UFO) in Austria, and other waste oils and fats in the United States. A 'new' diesel is also being made from coal and natural gas via the Fischer–Tropsch process, and in the future it may be possible to obtain it from lingo-cellulosic biomass using the same process (see, for example, www.ott.doe.gov/biofuels/).

Biodiesel can be used neat (100 per cent or B100) or in various blends. It can be used in any diesel engine with little or no modification to the engine or fuel system and does not require new refuelling infrastructure. Biodiesel maintains the same payload capacity and range as conventional diesel and provides similar or slightly reduced horsepower, torque and fuel economy.

Biogas

Biogas, a mixture of methane and carbon dioxide, is the most important gaseous biomass fuel. Biogas is produced in a digester by anaerobic bacteria acting on a mixture of dung and other vegetable matter mixed with water. The supply of biogas can be estimated by counting the number and capacity of digesters in a district or region, noting how many of them are actually functioning, and any variations in production through the annual cycle. A quick count of digesters can be made if necessary and conversion factors applied to give an estimate of production.

Biogas production and use can be grouped into three main categories:

- small domestic production/applications;
- small cottage industrial applications;
- industrial production/uses.

A significant change in biogas technology, particularly in the case of larger industrial plants, has been a shift away from energy alone towards more 'environmentally sound technology', which allows the combination of waste disposal with energy and fertilizer production, in both developed and developing countries. This has been helped by financial incentives, energy efficiency advances, dissemination of the technology and the training of personnel (Rosillo-Calle, 2006).

Various countries have large biogas programmes (e.g. China, Denmark, India, etc.), and many more are emerging as more and more countries consider the production of biogas from MSW, landfills, etc., viable for reasons stated above, although not on the scale of China and India where there are millions of biogas plants. Biogas, which was produced primarily on a non-commercial basis (e.g. for cooking, heating and lighting), is now also produced and used in large commercial plants around the world (e.g. for transport, heating and electricity generation, etc.), although mostly on an experimental scale.

One advantage of biogas is that it can use existing natural gas distribution systems and can be used in all energy applications designed for natural gas. However, a major disadvantage is its low calorific value; currently, one of the most widely uses is in IC engines to generate electricity.

Biogas is also compressed for use in light and heavy-duty vehicles. However, only a few thousand vehicles are thought to be using biogas in comparison to over one million using compressed natural gas (CNG). There are many experimental programmes around the world. In the past decade important breakthroughs for biogas production and technologies have been made in many countries, particularly in industrial nations, that will allow greater use of biogas in modern applications. However, the main driving force in biogas production is not energy, but the necessity of addressing environmental and sanitary problems. Thus biogas, rather than an alternative energy source, should be considered even more as a potential solution to environmental problems posed by excess manure handling, water pollution, etc. Future biogas applications will be primarily in modern rather than traditional uses.

Producer gas

Producer gas is generated by pyrolysis and partial oxidation of biomass resources such as wood, charcoal, coal, peat and agricultural residues. It is a practical and proven fuel that has many applications, for example, as a transportation fuel, as a boiler fuel for steam and electricity generators and other industrial uses. This technology is receiving increasing attention as it can prove particularly useful in many rural areas of developing countries, both as fuel for power generation (particularly co-generation) and heat production. However, production and running costs of producer gas continue to be high and applications would make sense only in niche markets and thus will not be considered in any more detail in this handbook.

Ethanol

Ethanol is the most important short- to medium-term alternative to petrol. In 2004/05 the two largest world producers were Brazil and the United States with over 15 and 14 billion litres respectively. Bearing in mind the current worldwide interest, it is expected that world production could reach 60–70 billion litres by the end of the decade.

Ethanol can be produced from any sugar-containing material and is currently produced from more than 30 feedstocks. However, in practice only a handful of raw materials are economically viable or close to being so. The principal feedstocks are sugar cane (Brazil) or molasses, and starch crops such as corn as in the United States. In the future one of the most promising feedstocks is cellulose-containing material, if production costs can be significantly reduced. There is abundant literature on this subject (see Chapter 1).

There are various types of ethanol as follows:

- biological or synthetic, depending on the feedstock used;
- anhydrous or hydrous, and denatured or non-denatured;
- industrial, fuel or potable, depending upon its use.

There are many sources from which ethanol can be produced although, as indicated, only a few are commercial or near-commercially viable. There are two main origins:

- biological – derived from, for example, grains, molasses, fruits, etc. (any sugar-containing materials);
- synthetic – derived from, for example, crude oil, gas or coal.

Although ethanol can be produced from very different types of raw materials, chemically the ethanol produced is identical. Also, be aware that there are two types of ethanol fuel:

- *hydrous* – meaning water-containing ethanol (usually 2–5 per cent water). This ethanol is used in neat ethanol engines (engines adapted to use 100 per cent ethanol). This is possible because, unlike when blended with gasoline, there is no phase separation. Only one country, Brazil, has produced large-scale neat ethanol vehicles; however, due to a combination of reasons (for example, the introduction of the multi-fuel engine) the neat ethanol car is being phased out.
- *anhydrous* – meaning water-free or absolute ethanol, which is blended with gasoline in different proportions.

And that ethanol can be:

- *denaturized ethanol* used as fuel – this is achieved by adding a small percentage of foreign material (which can be gasoline or other chemicals) that it is difficult to remove, to make it undrinkable.
- non-denatured ethanol or 'potable alcohol' which is the ethanol contained in beverages; it is the starting material used in the preparation of many industrial organic chemicals.

See Appendix VI Measuring Sugar and Ethanol Yields.

Methanol

Methanol (CH_3OH) is currently produced mostly from natural gas and also from coal, but recently there has been considerable interest in its production from biomass. Methanol from biomass requires pre-treatment of the feedstock, its conversion to syngas, which then has to be cleaned before being converted into methanol. This further increases production costs. Using this process, methanol can be obtained from the distillation of hardwoods at high pressure and a temperature of about 250°C, which is in itself a highly energy-intensive process.

Methanol is a common industrial chemical that has been commercialized for over 350 years and is widely used, primarily as a building block for thousand of products ranging from plastics to construction materials. Methanol has also been used as an alternative transport fuel blended in various proportions in many countries and is currently under consideration for wider use. Its main appeal is as a potential clean-burning fuel suitable for gas turbines and IC engines, but particularly for new fuel cell technologies (www.methanol.org/fuelcells/). The American Methanol Institute (AMI) estimates that methanol demand for fuel in the USA alone could reach over 3.3+ billion litres by 2010 (see www.methanol.org/).

However, it is the potential role of methanol to provide the hydrocarbon necessary to power fuel cells (FCVs) that currently makes it attractive. In 2000, the world production capacity of methanol was just under 47.5 billion litres. However, since today's economies favour natural gas, methanol from biomass remains a distant possibility, until it can be produced from other sources more competitively.

Methanol also offers few advantages over natural gas, except that it is a liquid and it is easier to use in a car. But the energy loss in the conversion of methane to methanol results in lower overall efficiency and higher overall CO_2 emissions than natural gas when used directly as fuel. In addition, the high toxicity of methanol makes it less attractive as a motor fuel. Thus, on a worldwide basis, the methanol market will remain relatively small, mainly for specialized markets such as chemicals and fuel cells, unless oil prices rise considerably.

Hydrogen

Hydrogen has been used for transport since the late 17th century.[3] However, it was not until the 1920s and 1930s that hydrogen was seriously considered as a transport fuel. In fact, vehicles using pure hydrogen have been designed and built for over 50 years (see, for example, www.e-sources.com/hydrogen/transport.html).

In recent years hydrogen has been researched in many countries as a potential fuel for transport. Hydrogen is believed by some experts to be a major fuel source for transport in the future, with its main potential in vehicles powered by fuel cells, although it is also a perfect fuel for the conventional gasoline engine. Hydrogen is not an energy source but an energy carrier and thus requires sources of energy, in exactly the same way as the other major energy carriers such as electricity. Like electricity, the advantage of using hydrogen as fuel, as far as security of supply or GHG emissions is concerned, depends on how hydrogen is produced.

Despite the large potential of hydrogen as a motor fuel, it seems obvious that the advantages of hydrogen will only be achieved after further successful technological development in storage, transportation, fuel cell technology, and production and distribution facilities. This will require costly investment that would have to be balanced against other alternatives with equal potential, such as biofuels, natural gas, etc.

DENSIFIED BIOMASS: PELLETS AND BRIQUETTES

Densification of biomass (pellets and briquettes) is a means of changing waste or low-value biomass products into products that can be used as fuel principally by industry or households, although there are other non-energy applications, e.g. flooring, furniture, etc. An important factor is that densified biomass is used in modern uses in both small- and large-scale applications. The most important material used includes agro-forestry residues such as sawdust, wood shavings, rice husks, charcoal powder or fines, sugar cane bagasse and also tall grasses. This process not only facilitates their use as energy but also reduces transport costs.

Densified biomass fuels are divided into two main categories:

- *pellets* – cylindrical pieces of compressed biomass, usually with a maximum diameter of 25 mm;
- *briquettes* – which can be cylindrical or any other form, also with a maximum diameter of 25 mm.

Generally, the size of pellets and briquettes varies from 10 to 30 mm, depending on its end use. Moisture content varies between 7 and 15 per cent. Densities can reach 1100 to 1300 kg/m^3, depending on the machinery used in the processing. Both pellets and briquettes can also be grouped into different categories depending on final use, with the higher quality Group 1 corresponding to small-scale users.

Densified biomass is acquiring increasing importance because of the growing domestic and industrial applications for heating, CHP and electricity generation in many countries. In countries such as Austria, Denmark, the Netherlands and Sweden, for example, it is becoming a major industry with pellets traded internationally. In Austria, the production of pellets in 2002 was 150,000 tonnes, but with the rapid expansion of small-scale pellet heating systems it is expected to reach about 0.9 Mt/yr by 2010 (http://bios-bioenergy.at/bios01/pellets/). Europe-wide this potential has been estimated at around 200 Mt/yr,[4] and is increasing continuously because advances in technology allow the densification of biomass to be more competitive, driven by high demand. The demand is for both domestic and industrial units in many industrial countries but also in many developing countries, particularly China. Thus, it is expected that this market will expand rapidly and even become an internationally widely traded commodity despite the growing importance of wood chips due to their lower cost.

Appendix 4.2 gives further information on densified biomass and there is a large amount of literature available on a growing number of Internet-based databases (see, for example, www.pelletheat.org; www.pellets2002.com/index.htm – this will take you to various other databases (European producers of densified biomass, retailers of pellets and briquettes, appliances, equipment, etc.) (www.sh.slu.se/indebif/; http://bios-bioenergy.at/bios01/pellets/en/; www.pelletcentre.info/)).

ANIMAL DRAUGHT POWER

Work performed by animals and humans is the fundamental source of power in many developing countries for agriculture and small-scale industries. Animals and humans provide pack and draught power, carry headloads and transport materials by bicycle, boat and cart. In the industrial countries, the work formerly done by human and animal draught power is now mostly carried out by all kinds of machinery. Further details are given in Appendix 4.3.

Draught animals are an important source of power for millions of people. There are about 400 million draught animals in developing countries that provide about 150 million HP annually. To replace this with petroleum-based fuels would cost hundreds of billions of dollars. Animal power varies considerably, for example in most African countries about 80–90 per cent of the population depends on manual and draught power, while the benefits of mechanization serve barely 10–20 per cent. In India between 50 and 60 per cent of the energy used in the agricultural sector is provided by draught animal power.

Thus, although animal draught power is gradually being replaced by machinery, it still constitutes a major source of motive power in the rural areas. Draught animals still offer advantages over mechanical draught power. For example, animals:

- have less impact on the land;
- can work in hilly and inaccessible areas, are 'fuelled' with feed, which can be produced locally, thus increasing fuel independence of the farmer;
- produce manure as a waste product, which is a valuable fertilizer;
- are cheaper to purchase and maintain than heavy machinery;
- are the sources of many other products (i.e. milk, meat, etc.).

One of the disadvantages of draught animals is that they need land on which to keep and produce their feed. This is a problem for many very small farmers. All kind of animals have been used for draught power, for example, horses, bulls and even dogs (see http://en.wikipedia.org/viki/Animal_traction).

Access to draught animals gives poor farmers in developing countries the means to accomplish the most power-demanding farm tasks: primary land preparation such as ploughing, harrowing, transport, and many industrial operations such as crushing and grinding. The alternatives to draught power are often tractors or other equipment, which in many developing countries are too expensive for small farmers, artisans, etc. Another alternative may be manual labour, which may restrict farm productivity and may be unpleasant and arduous.

Draught animal power should not be seen as a backward technology, in conflict with mechanization and modernization. On the contrary, given the widespread use of draught power and the fact that it will be the only feasible and appropriate form of energy for large groups of people for the foreseeable future in many countries, it should be given prime attention.

The working performance of a draught animal is primarily a function of its weight, provided it consists of muscle rather than fat. For example, oxen can be expected to exert a draught force of about 10 per cent of their body weight and horses about 15 per cent. In terms of efficiency, the majority of draught animals produce 0.4–0.8 HP on a sustained basis. However, poor dietary levels often result in draught animals delivering far less power than they would be capable of with proper feeding. An increase in the productivity of the estimated 400 million draught animals in use may possibly raise the productivity of small farms. It is important that farmers recognize this potential for improvement in performance and actively seek increased draught animal productivity.

Performance in field conditions of both mechanical and physiological parameters can be measured with varying degrees of sophistication. The draught force can be measured with a spring balance, hydraulic load cell or electronic strain gauge cells. The distance travelled is simply measured with a tape measure and speed with a stop watch. Farmers are adept at judging the degree of fatigue of their draught animals. Nevertheless, some researchers have suggested a more sophisticated system based on respiration rate, heart rate, rectal temperature, leg coordination, excitement, etc. (see Appendix 4.3 for further details).

Future options

Residues (all forms) are bound to increase in importance for various reasons:

- they are currently an underused and undervalue resource;
- they are becoming economically and environmentally more attractive;
- they are, in most cases, a readily available alternative.

Greater understanding of ecological issues, current and future energy potential and the economics of using agricultural residues is long overdue, and thus methodological assessment improvements will occur.

Animal residues (particularly manure) are likely to play a diminishing role in a modern world. It is likely that the large-scale use these residues as a source of energy will be justified only in specific circumstances, for example in locations where a combination of environmental, health and energy use are important. The methods for assessing manure suffer from many shortcomings due to large variations in animal size, feeding methods, etc.

The most promising are tertiary residues (e.g. MSW) which are increasingly becoming an attractive option around the world and for which more refined assessment methods will be required.

Herbaceous energy crops are bound to play a significant role partly because they can be harvested annually and are very attractive to the farmer who can use basically the same machinery as for food crops. In this case, modern agricultural methods will apply.

Secondary fuels (ethanol, biodiesel, etc.) are already used on a large scale in some countries. However, much needs to be learned about the potential effects of the methods of their large-scale production and use. Large-scale development of such fuels will have major impacts on biomass resources.

Densified biomass, as with secondary fuels, is already being used as a major source of energy, primarily in some European countries, although an internationally agreed standard measurement method is still lacking.

Finally, animal power – so vital in the past – is becoming less and less important as many of their activities are being replaced by machinery. However, draught animal power offers many advantages and should not be overlooked, but this is a complex issue beyond the scope of this handbook.

NOTES

1   See Ryan and Openshaw (1991) and Openshaw (1998).
2   This compares with a planted area of 25.5 Mha in 2005.
3   Hydrogen was used by the Montgolfier brothers to power their balloons.
4   In practice, this potential will be severely limited due to high costs.

REFERENCES AND FURTHER READING

Anon (1988) *Wood Densification*, West Virginia University, Extension Service, Publication no 838

Bullard, M. and Metcalfe, P. (2001) *Estimating the Energy Requirements and CO$_2$ Emissions from Production of the Perennial Grasses Miscanthus, Switchgrass and Reed Canary Grass*, ETSU B/U1/00645/REP; DTI/Pub., URN 01/797.

Hirsmark, J. (2002) 'Densified Biomass Fuels in Sweden; Country Report for the EU/ INDEBIF Project', Sverges Lantbruks Universitet

Horta, L. A. and Silva Lora E. E. (2002) 'Wood Energy: Principles and Applications (unpublished report)', Federal University of Itajuba, Minas Gerais, Brazil

Kristoferson, L. A. and Bokalders, V. (1991) *Renewable Energy Technologies*, Intermediate Technology Publications, London

Larson, E. D. and Kartha, S. (2000) 'Expanding Roles for Modernized Biomass Energy', *Energy for Sustainable Development*, vol 5, no 3, 15–25

Moreira, J. R. (2002) 'The Brazilian Energy Initiative: Biomass Contribution', Paper presented at the Bio-Trade Workshop, Amsterdam, 9–10 September 2002

Obernberger, I. and Thek, G. (2004) 'Physical characteristics and chemical composition of densified biomass fuels with regard to their combustion behaviour', *Biomass and Bioenergy*, 27, 653–669

Openshaw, K. (1998) 'Estimating biomass supply: Focus on Africa', in *Proc. Biomass Energy: Data Analysis and Trends*, IEA/OECD, Paris, pp241–254

Rogner et al (2001) 'Energy resources', in *World Energy Assessment: Energy and the Challenge of Sustainability; Part II Energy Resources and Technology Options*, Chapter 5, UNDP, pp135–171

Rosillo-Calle, F. (2004) *Biomass Energy (Other than Wood)*, World Energy Council, London, Chapter 10, pp267–275

Rosillo-Calle, F. (2006) 'Biomass energy', in Landolf-Bornstein Handbook, vol 3, *Renewable Energy*, Chapter 5 (forthcoming)

Ryan, P. and Openshaw, K. (1991) *Assessment of Biomass Energy Resources – A Discussion on its Need and Methodology*, Industry and Energy Dept. Working Paper, The World Bank, Washington DC

Sims, B. G., O'Neil, D. H. and Howell, P. J. (1990) 'Improvement of draught animal productivity in developing countries', in *Energy and the Environment into the 1990s*, Sayigh, A. A. M. (ed), Pergamon Press, vol 3, pp2958–1964

Starkey, P. and Kaumbutho, P. (eds) (1999) *Meeting the Challenges of Animal Traction. A Resource Book of Animal Traction*, Network for Eastern and Southern Africa (ATNESA), Harare, Zimbabwe; Intermediate Technology Publications, London, 326pp

Woods, J. and Hall, D. O. (1994) *Bioenergy for Development: Technical and Environ-mental Dimensions*, FAO Environment and Energy Paper 13, FAO, Rome

Websites

http://bios-bioenergy.at/bios01/pellets/
http://bios-bioenergy.at/bios01/pellets/en/
http://en.wikipedia.org/wiki/Animal_traction
http://www.esv.or.at/
www.e-sources.com/hydrogen/transport.html
www.methanol.org/
www.methanol.org/fuelcells/
www.pelletheat.org
www.pellets2002.com/index.htm
www.ruralheritage.com/horse_paddock/horsepower.htm
www.sh.slu.se/indebif/
www.worldwideflood.com/ark/technology/animal_power.htm

APPENDIX 4.1 SUGAR CANE BAGASSE ENERGY COGENERATION: THE CASE OF MAURITIUS SOURCE: DEEPCHAND (2003)

Sugar cane residues (bagasse, tops and leaves) are one of the most promising energy alternatives (see Figure 4.1). Almost all sugar cane mills around the world

Figure 4.1 *The component fractions of sugar cane*

are self-sufficient in energy just by using bagasse, even when used at very low efficiency. In recent years many mills have been or are being modernized, to take greater advantage of the energy potential through co-generation, e.g. Brazil has more than 1.5 GW installed capacity, offering the opportunity to sell energy surpluses to the grid. Despite the small area planted to sugar cane in the world of approximately 25 Mha,[1] compared to over 250 Mha of wheat, its energy potential is large. Many studies have been carried out to assess this potential although there are large discrepancies.

For example, Moreira (2002) has estimated that 143 Mha of new sugar cane plantations worldwide (about 5.7-fold increase in planted area) could generate 47.36 EJ (26 Mboe/day of ethanol and 10,000 TWh/yr of electricity) by 2020. Although this is a very optimistic estimate, it is feasible, since sugar cane is produced in 102 countries and large productivity increases are possible with improved management and without any major investment. This assumes that new technologies and modern management practices will be applied throughout in response to pressure to modernize and diversify and to find alternative uses for sugar cane and by-products. This appendix describes co-generation in the sugar cane industry in Mauritius.

The island of Mauritius has an area of 1860 km² and a population of 1.3 million. Sugar production has increased over time to reach a plateau of around 600,000 to 650,000 t. The two main limitations to the cultivation of sugar cane are the availability of arable land and export difficulties. Three distinct groups of growers own the total area under cane:

- the miller-planters, with a majority share in the milling companies which hold 55 per cent of the cane area;
- individually owned small size varying between 700 to 5500 ha, responsible for 60 per cent of total sugar;
- about 35,000 independent growers holding approximately 200,000 plots whose areas vary between 0.1 to over 400 hectares in same class.

The independent growers and the miller-planters are entitled to 78 per cent of their sugar and the totality of the molasses and filter mud. The millers obtain 22 per cent of the sugar as payment for milling.

Energy status

Mauritius has limited renewable energy resources and no known oil, gas or coal reserves. Its main locally available energy resources are hydropower and sugar cane biomass. Hydropower is almost fully exploited with its nine hydro stations including one with 10 MW installed capacity. The other resource, sugar cane bagasse, which represents 30 per cent of the cane, was generally being used inefficiently to meet internal power requirement for cane processing.

Hydropower and power exported to the grid from sugar factories amounted to 22 per cent and 13 per cent of power supply respectively to the public grid in the year 1990. The remaining 65 per cent was met from imported fossil fuels (diesel, coal and gas). It was felt that a rapid increase in fossil fuel import could be prevented by a more efficient exploitation of bagasse energy for electricity generation.

Energy demand in Mauritius has increased sharply in the past 20 years. Of the various options available to meet the increase in demand of electricity, the government chose to purchase the bulk of power from two 22 MW bagasse-cum-coal plants, to be privately operated by sugar companies at two regional sugar factories.

Objectives of bagasse energy development

A bagasse energy development programme (BEDP) was formulated by the government in partnership with the private sector over a six-month period in 1991 on the basis of the recommendations of the High Powered Committee on Bagasse Energy. The programme had two main objectives:

1   to optimize the use of bagasse for electricity generation and export to the grid. Over the five-year period to expand electricity generation using bagasse from 70 GWh to 120 GWh;
2   to investigate the use of other fractions of the sugar cane biomass (cane tops, leaves and dry trash) for electricity generation in order to further reduce dependence on fossil fuel.

An important aim was to ensure the continued viability of the sugar sector and sustainability of production to meet the industry's commitment under the preferential sugar market. The project required investment of US$80 million (1991 prices) in the following:

•   building and commissioning bagasse-cum-coal fired power plants at two sugar factories;
•   modernization of sugar factories to improve the efficiency of bagasse use in sugar cane processing;
•   bagasse transport from a cluster of sugar factories to a regional sugar factory located power plant;
•   investment in transmission lines from sugar factories to the national grid.

Institutional set-up and project strategies

In the implementation of the project, a regulatory framework was set up to promote private sector investments in power production and sugar factory modernization and to encourage an efficient market in bagasse. The key element of this framework was energy pricing and contracting, involving electricity, bagasse and coal.

A Management Committee with representatives from government and industry was set up to undertake detailed planning of programme implementation, to ensure that the government's policy directives related to the BEDP were followed and to effectively integrate government policies affecting the sugar and energy sectors. All the relevant parties (ministries and agencies, the public utility, the private sugar industry stakeholders) fully participated in the project from inception through all stages.

Implementation of bagasse energy projects

The project duration was five years with the effective start in 1994. However, it was envisaged that the use of bagasse would be optimized by the year 2000. The project stages were as follows:

1 Government policy established defining clearly the bagasse energy option as a means to promote a renewable energy resource available locally.
2 Sugar industry to evaluate its energy requirement and optimization of same through proper investments in measures for energy conservation and use.
3 Public utility to spell out its energy demand based on reliable forecasts in order to establish its base load requirement over time.
4 Memorandum of understanding between the utility and the sugar company drawn up.
5 A feasibility study conducted.
6 Signing of a formal power purchase agreement (PPA) between utility and the private investor.
7 Raising of funds for investment in power plant using PPA as the bank guarantee.
8 Carrying out a detailed design of project.
9 Carrying out a tendering exercise for supply of items of equipment.
10 Evaluation of tenders.
11 Award of contract.
12 Erection and commissioning.
13 Operation.

Constraints to bagasse energy development

In spite of all the above measures, it was observed that investment in bagasse saving in the satellite factories was slow and only 40 per cent of the total amount (US$15 million) of the Sugar Energy Development Plan Loan was disbursed. Several factors were identified which had influenced this state of affairs.

Price of bagasse

The progress in the implementation of the power plant at the Union St Aubin sugar factory was slow because the plant had to rely on a huge amount of bagasse from the satellite factories. These factories were costing the bagasse based on the price of coal and at the condensation mode of operation, during which the efficiency of conversion of steam into electricity is higher compared to that of a condensation-extraction mode of operation, which is the usual industrial set-up for energy co-generation in the sugar sector. This price had a negative impact on the financial viability of the project. This issue was resolved through consolidation of cane milling activities whereby the cane was processed in an increasingly smaller number of sugar factories whose cane-crushing capacity was increased, and which invested in power plants.

Funding and the fiscal framework

The energy projects required a relatively huge investment cost that made it unattractive. Hence, the government introduced several enactments which allowed investors to raise tax-free debentures for the generation of electricity from bagasse and the modernization of sugar factories. This enabled growing companies with segregated activities to offset losses incurred by millers in respect of the capital expenditure in energy production from bagasse and in the modernization of sugar mills. Furthermore, the performance-linked rebate on export duty was extended to producers of electricity who saved and used their own bagasse, and also to millers selling bagasse to power stations that were operating continuously. A proportion of capital expenditure incurred in the installation of efficient equipment that used less bagasse while at the same time saving energy was therefore entitled to a refund of export duty.

Any bagasse used for purposes other than the manufacture of sugar was priced at Rs100 (or US$3.7) per tonne, and most of the monies raised were credited by the Central Electricity Board (CEB) to a bagasse transfer price fund. The distribution of the proceeds from that fund was modified so that millers or sugar factory based companies exporting electricity to the CEB became entitled to benefit from the fund. This fund had previously been accruing to growers only.

Centralization of cane-milling activities

Consolidation of cane-milling activities through centralization is one means of reducing costs of production. A total of 19 sugar factories were in operation in 1993 and their cane-crushing capacity ranged between 55 to 250 tonnes cane hour (tch). In 1997, the government came up with a blueprint for the Centralization of Cane-Milling Activities which, besides setting guidelines and conditions to be adhered to, in any request and implementation of such closures, emphasized the need to link such closures with energy generation from bagasse.

The kWh price

The government set up a Technical Committee to address the issue of energy pricing and power purchase agreement. In the price-setting mechanism, the Committees worked on the basis of the cost of a diesel plant of 22 MW capacity proposed by the CEB to arrive at the avoided cost for the power plant. The World Bank provided support to the Committee to work out the principles and the guidelines. This Committee determined the avoided costs and recommended the kWh price for coal and bagasse.

Evaluation of project implementation

The activities related to the project were undertaken as planned but there was a delay in its date of completion, mainly due to the fact that the investors in the Union St Aubin plant decided not to go ahead with their project due to high costs. In 1995 the design firm recommended redesigning major components such as the boiler and the turbo-alternator, to take account of future capacity of the factory, and improvements in the thermodynamic cycle of the plant. This new design brought about a 30 per cent increase in cost of the previous 30 MW plant design. Under the circumstances, the overseas bank which was interested in funding the foreign exchange decided not to fund the project.

Almost immediately afterwards, Centrale Thermique de Belle Vue, having learned from the experience and studies undertaken at Union St Aubin, started negotiations for the building of a 70 MW plant (two units of 35 MW each) and commissioned the power plant in April 2000. This plant required an investment of US$90 million.

Bagasse energy projects

Table 4.2 shows the status of the energy projects and it includes technical details on the ten bagasse-based power plants. Three of the power plants operate year-round, using bagasse during crop season and coal during the off-crop period. The so-called 'continuous' power plants operate during the crop season only and use bagasse as the combustible medium.

Progress on bagasse energy evolution

The outcome of the project has been satisfactory in that its key strategy was to set up an investment plan, the institutional framework and the policies to encourage private investment in bagasse/coal power plants. This was achieved under the project and more bagasse units have been projected. As at the year 2000, the

Table 4.2 *Bagasse-based power plants in Mauritius up to the year 2000*

Factory	Tonnes cane per hour	Power	Start date	Units from bagasse (GWh)	Units from coal (GWh)	Total units from bagasse and coal (GWh)
FUEL Deep River	270	F	October 1998	60	115	175
Beau Champ	270	F	April 1998	70	85	155
Belle Vue	210	F	April 2000	105	220	325
Médine Mon Tresor	190	C	1980	20	–	20
Mon Desert Union	105	C	July 1998	14	–	14
St Aubin	150	C	July 1997	16	–	16
Riche en Eau	130	C	July 1998	17	–	17
Savannah	135	C	July 1998	20	–	20
Mon Loisir Mon Desert	165	C	July 1998	20	–	20
Alma	170	C	Nov 1997	18	–	18
Total		**3 F** **7 C**		**360 GWh** **235 GWh F** **125 GWh C**	**420 GWh**	**780 GWh**

Notes: F = firm or bagasse during crop season and coal during intercrop periods.
C = continuous or bagasse during crop season only.

bagasse-cum-coal power plants accounted for 220 MW installed capacity or more than 50 per cent of the total (425 MW). Two additional firm power plant projects have already been formulated and were awaiting an audit which will establish the power demand and, more particularly, evolution of base load over the next decade prior to implementation. The project related to investments in mill efficiency had a positive outcome in that significant improvements were made in energy use and conservation in cane processing. The tangible result on this project is in the amount of surplus bagasse generated from the sugar factories.

In 1996, 119 GWh of electricity was exported from bagasse, which is almost the 120 GWh target specified in the project objective. This was achieved mostly through investment by private sugar mills using co-generation technology with their own private fund. By the year 2000, co-generated energy increased significantly with investment in more efficient bagasse-to-electricity processes and in a greater number of units. As a result, the electricity exported to the grid from bagasse increased to 274 GWh from the 160 MW installed (or 33 per cent) firm installed capacity.

Factors inducing bagasse energy development

Bagasse energy projects are linked with sugar factory modernization in that boilers, turbo alternators and other energy-efficient equipment represent a major proportion (up to 50 per cent) of the cost of a sugar factory. Investing in an energy project ensures that this part of the investment (useful life of 25 years) crucial to sugar processing, is financed independently of sugar activities. In addition, the sale of electricity adds to the revenue of the sugar companies. In 1985, 21 sugar factories were in operation and the number has decreased to 14 in the year 2000. Ten of these factories exported energy to the grid and only three of them were firm power plants. It has been projected that by the year 2005, only seven sugar factories will be in operation through the process of centralization and it is probable that each one of them will be equipped with a firm power plant, which are generally more efficient in energy co-generation and so export more power to the grid.

Replication opportunities and sustainability of bagasse energy

With the successful demonstration of the bagasse energy projects in Mauritius, opportunities are now offered to other cane sugar producing countries to replicate or adapt such projects.

The kWh/t cane processed in 1988 was 13, and even after implementation of the projects by the year 2000, the value had reached just 60 kWh/t cane. This is well below the 110 kWh/t cane obtained in Réunion where only two factories were in operation, equipped with 2×30–35 MW power plants operating at around 82 bars. With further centralization of cane-milling activities, improvement in exhaust steam in cane processing, upgrading the efficiency of the power plants by adopting operating pressures of 82 bars and use of cane field residues as supplementary fuel, it can be safely said that 800 GWh of electricity can be exported to the grid from sugar cane biomass.

Lessons learned and recommendations

The main lesson learned is that development of bagasse-based electricity generation required stronger linkage between the sugar industry and those in the power sector, as well as a greater emphasis on multipurpose benefits resulting from the generation of base load power from bagasse/coal plants.

The government's strong support, clearly defining its policy with respect to bagasse energy development, is critical to the successful implementation of co-generation using bagasse.

Conditions must be created to enable all the stakeholders to participate fully in the whole process, as well as establishing transparent flow of information among them. In this case the World Bank played a key role in providing the necessary support in areas in which the local stakeholders had little or no experience.

Prior to start of a bagasse power plant development, it is of the utmost importance that a detailed feasibility study including a reliable cost estimate for a bagasse/coal plant and an agreement on a financing plan from the private entrepreneur are made available. This will avoid delays in project implementation.

The bagasse/coal power development has multiple benefits in that it is associated with environmental advantages, offers a diversified alternative and secure source of power from a locally available and renewable resource when compared to imported fuel oil and, finally, brings additional revenue to the cane sugar industry.

Note

1   By a large margin, the world's two largest producers are Brazil with 5.5 Mha and India with just over 4 Mha.

Appendix 4.2 Densified biomass: pellets

Growing interest in densification of biomass has led to a rapid improvement in compacting technologies which have been able to borrow from other industrial sectors, for example, fodder and oil distribution systems. See the references under 'Further reading' for follow-up information on these advances.

Easy to transport and to handle, wood pellets are the fuel of choice in some parts of Europe such as Austria and Germany where they tend to be used at a domestic level, and Scandinavia, where the pellets are use in CHP and district heating and more recently in cofiring plants.

The introduction of pellet use in large-scale boilers of up to 100 MW installed capacity was a major step forward in the consolidation of the pellet industry, partly driven by tax incentives. For example, in the early 1990s Sweden introduced a new tax on fossil fuel CO_2 emissions and this paved the way for a quick expansion of the pellets market.

Between 1992 and 2001, pellet consumption in Sweden increased from 5000 tonnes to 667,000 tonnes per year – making Sweden the largest pellet user and producer in Europe, with about 30 large production plants.

All types of woody biomass are in principle suitable raw materials for wood pellet production. However, to keep costs for drying and grinding low, dry sawdust and wood shavings are predominantly used. The production of pellets from bark, straw and crops is usually more appropriate for larger-scale systems. The requirements for pellet fuel quality vary according to the different characteristics of the wood pellet markets in different European countries. For example, Table 4.3 summarizes the characteristics required in Austria.

Table 4.3 *Criteria for Austrian standard ÖNORM M 7135*

Property	Pellets class HP1 (wood pellets)
Diameter	Min. 4 mm, max. 10 mm
Length	Max. 5 × diameter
Density	Max. 1.12 kg/dm^3
Water content	Max. 10%
Abrasion	2.3%
Ash	Max. 0.5%
Caloric value	Min. 18 MJ/kg
Sulphur	Max. 0.04%
Nitrogen	Max. 0.3%
Chlorine	Max. 0.02%
Additives	Max. 2%

The following are some additional points regarding raw materials for densification of biomass:

- Densified biomass can vary significantly and has different physical and chemical characteristics. For example, bark pellets have higher ash content than wood pellets and also higher emissions from combustion than wood. Bark, when burned in its raw form, is used in special boilers (e.g. in Sweden where it is very abundant). To mitigate some of these problems, bark may be mixed with wood.
- When tall grasses are used (e.g. elephant grass, switchgrass), they are densified into bales, although they might also be called briquettes. Bales can weigh 250–500 kg. Grasses are becoming increasingly attractive in certain areas which do not have other alternative energy sources and have high costs of conventional energy. Grasses are particularly attractive for co-firing with coal.
- Peat is also an organic matter that originates from biomass that does not decompose in the natural environment. Although opinions are divided when it comes to classifying it, either as biomass or fossil fuel, it is usually considered as non-biomass.
- Densified biomass is obtained in most cases from residues (excluding grasses). The use of energy plantations may only be justified if residues are not available.
- The raw material potentially available for densified biomass is enormous, but cost (energy spent, machinery, etc.) is a major limiting factor.
- Raw material potential for densification of biomass changes with improvements to existing technology and new technological developments.

The production of densified biomass fuels (DBFs) entails various steps:

- Moisture content is very important. It is also important that the material is as dry as possible. This means that some type of drying equipment will be needed in most cases. Material with high moisture content is very difficult to condense.

- Particles should be of a certain size, depending on the end use.
- Conditioning may be necessary, for example to make wood fibres softer and flexible by applying superheated steam to the raw material to facilitate sizing.
- The decision must be made regarding what to produce: pellets or briquettes, for example. Different machinery is needed, e.g. flat die pelletizers for wood and ring die pelletizers for bark. For example, in a ring die pelletizer the die is cylindrical and the pellets are pressed from the inside and out through the die by rollers. For briquetting, the dominant technology is the piston press that pushes the raw material through a narrow press cone by either a mechanically or hydraulically driven piston (see Obernberger and Thek, 2004).
- Cooling is necessary to reduce the steam pressure of pellets and briquettes and prevent them breaking up through vapour pressure.

Costs have been a major obstacle in the past in preventing the use of biomass densification on a large scale and are key factors to take into account. Costs, together with high energy consumption of the process, have improved significantly thanks to new technological advances in this field. These factors have also been assisted by higher oil prices and increasing environmental concerns.

Various studies have looked in considerable detail at the costs of densification, particularly in Austria, the USA, Canada, Sweden and the Baltic countries (see Zakrisson, 2002 – quoted in Hirsmark, 2002). The various costs involved in the densification of biomass can be divided into four main categories:

1 Costs based on capital and maintenance – these costs are computed using the calculated service lives of the equipments and interest rate. Capital costs are equal to the investment costs multiplied by the capital recovery factor (CRF). Maintenance costs are percentages of the investment costs, and are calculated on the basis of guideline values.
2 Usage-based costs – which include all costs in connection with the manufacturing process.
3 Operational costs – which include all costs involved in operating the plant, such as personnel costs.
4 Other costs – this item includes insurance, taxes and administration.

Environmental impacts of DBFs

The environmental impacts of DBFs also need detailed consideration when dealing with biomass densification, particularly in the case of a medium or, even more importantly, a large-scale plant. Various life cycle analysis (LCA) studies have been carried out on densification of biomass, particularly in the Scandinavian countries, for both heating and electricity which will provide invaluable information. Existing data must be assessed to determine if it is adequate or whether a new LCA is needed.[1]

Standards

International standards can vary significantly and this is obviously a barrier for international trade. It is important to develop some internationally agreed standard for densified biomass. The Austrian standard ÖNORM M 7135 ensures a high quality of compressed biomass fuels and the exclusive use of natural raw materials. The main criteria are shown in Table 4.3.

Note

1   Christiane Egger, Christine Öhlinger and Dr Gerhard Dell work for O. Ö. Energiesparverband, the regional energy agency of Upper Austria.

Further reading

Hirsmark, J. (2002) Densified Biomass Fuels in Sweden; Country Report for the EU/INDEBIF Project, Sverges Lantbruks Universitet

Obernberger, I. and Thek, G. (2004) 'Physical characteristics and chemical composition of densified biomass fuels with regard to their combustion behaviour', *Biomass and Bioenergy*, 27, 653–669.

www.esv.or.at/

www.pelletcentre.info/

APPENDIX 4.3 MEASURING ANIMAL DRAUGHT POWER

Two instrumentation packages have been developed to monitor draught animal performance under field conditions: an ergometer which measures mechanical and some physiological variables, and the draught animal power logger designed to measure simultaneously both mechanical and physiological parameters so that the animals' responses to varying field working conditions can be measured.

There are two main categories of variables:

1   mechanical variables, e.g. vertical and horizontal components of draught force and speed, and
2   physiological variables, e.g. oxygen consumption, heart rate, breathing rate, etc.

Attempts to measure these variables have been carried out in both simulated and actual field conditions, for example loaded sledges pulled by animals via a force transducer, comparison of yokes and harnesses.

Comparing power from animals (oxen) with energy from wood

It is not possible to compare power and energy. A comparison can be made only by specifying a time period (for example, the time period of an ox's work). Typically, a good ox can deliver 0.8 HP, about 600 W. The amount of energy delivered in one year by this type of ox can easily be calculated if one knows the number of hours an ox works per day and how many days per year. For example, if an ox works five hours per day and 280 days per year, the energy from an ox per year will be:

$$600 \text{ W} \times 1400 \text{ hrs} \times 3600 \text{ W} = 3.0 \text{ GJ (about 1/6 of a tonne of wood)}$$

Comparing human heat and energy

A human body doing little or no physical work needs about 2000 kcal of energy in its daily diet which is converted into heat. Thus:

$$2000 \text{ kcal/day} = 2000 \times 4.2 \text{ kcal/day} = \frac{8.4}{86{,}000 \text{ s/d}} = 100 \text{ J/s} = 100 \text{ W}$$

Therefore, a human body doing little or no work generates a heat equivalent to 100 W.

The daily energy in a person's diet is 8.4 MJ. If we assume that the food mainly consists of crop products, i.e. biomass, and considering that dry biomass has an energy content of about 18 MJ/kg, then:

$$\frac{8.4 \text{ MJ/day}}{18 \text{ MJ/kg}} = 0.5 \text{ kg/day of biomass for food}$$

On a yearly basis, the biomass for food per person is:

$$365 \times 0.5 \text{ kg/day} = 180 \text{ kg/yr}$$

From surveys, we know that household fuel needed for cooking is about 500 kg/yr of dry biomass per person. Hence, the ratio of the fuel to food is:

$$500 = \text{approx. } 2.7 \times 180$$

This means that nearly three times as much energy is required under the pot as in the pot!

Further reading

Deepchand, K. (2003) 'Case Study on Sugar Case Bagasse Energy Cogeneration in Mauritius', Dept. of Chemical and Sugar Engineering, University of Mauritius, Reduit, Mauritius, www.iccept.ic.ac.uk/research/projects/SOPA/PDFs/Energy%20Fiji.pdf

Kristoferson, L. A. and Bokalders, V. (1991) *Renewable Energy Technologies*, Intermediate Technology Publications, London, pp119–132

Sims, B. G., O'Neil, D. H. and Howell, P. J. (1990) 'Improvement of draught animal productivity in developing countries', in *Energy and the Environment into the 1990s*, Sayigh, A. A. M. (ed), Pergamon Press, vol 3, pp2958–1964

Animal traction see http://en.wikipedia.org/wiki/Animal_traction

www.ruralheritage.com/horse_paddock/horsepower.htm

Lover Mar, T. (2004) *Animal Power* available online at (www.worldwideflood.com/ark/technology/animal_power.htm)

5

The Assessment of Biomass Consumption

Sarah L. Hemstock

INTRODUCTION

This chapter deals in some detail with various methods for obtaining reliable data on biomass energy consumption. It is structured to have particular relevance for the field worker looking at the feasibility of smaller scale bioenergy projects. The emphasis is on community level consumption in rural areas of developing countries, with respect to the amount and type of biomass resource consumed and that available for project activities. Sections of this chapter examine suitable assessment methods, appropriate analysis and the assessment of availability of appropriate resources for satisfactory formulation of a bioenergy project. Indicators of changes in biomass consumption over time, which may alter the amount of biomass resource available for future project sustainability, are also examined. An example of a survey designed and implemented in the field (by the NGO Alofa Tuvalu)[1] to assess biomass consumption in Tuvalu will be used to illustrate some of the issues discussed in this chapter.

Initially, survey design and implementation will be examined since surveys are important tools in determining biomass consumption patterns. Analysis of domestic energy consumption will be investigated, as the domestic sector is responsible for a very large proportion of biomass consumption. The section on 'Analysing domestic consumption' provides information on how to measure and analyse biomass consumption and the section on 'Changing fuel consumption patterns' gives step-by-step details on how these patterns can change over time.

The flow chart methodology described in Chapter 2 is another useful means of calculating bioenergy consumption at various scales – local, national and regional. Analysis of the availability and consumption of the biomass resource is crucial if biomass energy is to be used on a sustainable basis. Calculating biomass energy flows – using reliable data – is one tool which allows consumption to be gauged and predictions to be made concerning availability, consumption and sustainability of biomass energy and also highlights areas in the harvest to end-use chain where improvements in efficiency can be made.

Designing a biomass energy consumption survey

Any survey will produce data that is biased, particularly if the data is gathered by questionnaire. Therefore, before embarking on an appraisal it is necessary to establish the objectives of the survey and determine what data is required from it. In particular, the following questions should be asked:

- What is the reason for the survey? Consider the advantages and disadvantages of using questionnaires in your particular circumstances.
- What questions need to be answered?
  — Prepare written objectives for the research.
  — Have your objectives reviewed by others.
- Is the survey strictly necessary?
  — Biomass energy use often fails to be included in official government statistics. However, it is worth reviewing any literature related to your objectives since surveys are time consuming and costly, so you do not want to repeat the work of others.
- What actions/interventions/activities are going to be based on the results of the survey?
- What level of detail is required?
- Which sectors will be surveyed?
- What are the resources available? (For example, if you decide on a questionnaire, determine the feasibility of administering it to the population of interest.)
- How long is this process going to take? Prepare a time line.

The answers to these questions may suggest a methodology for data collection, compilation and presentation. The essential details can then be expanded where it is necessary, desirable or possible.

Field example

Alofa Tuvalu – Tuvalu Biomass Energy Consumption Survey
The objective of the survey was to determine the amount of biomass available for two community bioenergy projects in Tuvalu (pig waste biogas digestion

and coconut oil biodiesel) (Hemstock, 2005). A detailed literature review was undertaken and it was found that biomass energy use was not included in any previous contemporary literature in the detail required. Other domestic energy use (such as liquid petroleum gas (LPG) and kerosene) had been estimated in a recent household survey based on the number and type of cooking appliances. This assessment was not accurate as actual fuel use was not accounted for. Literature detailing standing stock, vegetation type/class, land use issues and house construction was also reviewed. In order to establish the amount of biomass available for these two projects it was deemed necessary to carry out a detailed assessment of energy use in the domestic sector.

The results from this survey could then be considered against annual productivity and standing stock. The best method of carrying out the survey was by questionnaire and by weighing wood piles. The questionnaire survey method was advantageous since households were close together and it was possible for project staff to question each household individually so a fairly accurate assessment of domestic bioenergy use could be obtained. In addition, local organizational infrastructure was engaged and women were questioned at a local women's group and 'Kaupule' (local council) meetings. Language was one disadvantage that had to be overcome using local representatives. A survey team consisting of ten local representatives (nine of them women) and two project staff (both women) carried out the survey across three of Tuvalu's nine islands over a six-week period in 2005.

Questionnaire design

Very often, surprisingly few fairly general questions can capture many of the issues relating to biomass consumption (food and fodder use, construction, domestic energy, fibre for mats and clothing, fertilizer). Surveys/questionnaires are important tools in helping to form a realistic picture of biomass consumption patterns. You should focus initially on issues you consider absolutely necessary. It is important to underline the preliminary nature of any programme design at this stage. Do not forget that local conditions play an important role in the final shape of the survey.

In order to obtain reliable data you need to:

- Group the items by content, and provide a subtitle for each group.
- Within each group of items, place items with the same format together.
- Indicate what respondents should do next or the use to which the information will ultimately put at the end of the questionnaire.
- Prepare an informed consent form, if needed.
- Consider giving a token reward for completion of the questionnaire. (To encourage participation a gift of vegetable seeds for family gardens was given by Alofa Tuvalu personnel.)

- Consider preparing written instructions for administration if the questionnaire is to be administered in person (the preferred method).

Detailed village surveys are a crucial element in the preparation of biomass energy projects. They provide an opportunity for people to express their opinion about the problems they face and how best to solve them, and for local communities to be involved with survey design. It is important that local people are actively involved from the beginning, and that they trust and gain the confidence of those who pose the questions. It is also important to ask the right questions of the right people. For example, if the survey is dealing with household energy consumption then it is wise to ask the woman's point of view as she will be able to provide more accurate information, and any project intervention is likely to affect her more directly.

Outlining required information

Problems can occur when elements of the survey are not strictly defined. It is therefore important that interviewers and interviewees understand exactly what is meant by the survey questions and what analysis is required from the answers. For example, a question as simple as 'household size' is not as easy to define as it first appears, and may not relate to the number of people using domestic bioenergy for their cooking needs.

Example: defining a household

Domestic use often accounts for the largest consumption of biomass fuel in many developing countries. However, it is important to be clear about exactly what we mean by 'household' before considering how the sector should be disaggregated. The term 'household' is subject to various definitions, and estimates of per capita consumption will vary according to which one is used. At different times, operative household size may be defined as follows:

- the number of family members (or cooking unit) who sleep and eat in the house;
- the number of family members who sleep but do not eat in the house;
- the number of family members who eat but do not sleep in the house.

Definitions may also include:

- the number of labourers who eat but do not sleep in the house;
- the number of people who work elsewhere, but who regularly eat some meals with the household;
- the number of people naturally included as members of the household (e.g. head of household who is a labourer, but who actually eats elsewhere);

- the number of people for whom food is cooked and to whom it is taken, but who actually live elsewhere (e.g. grandparents and children).

There are yet further possibilities. For example, a working member of the family may receive payment in food that is brought home, rather than money. In this case it is important to discover whether the food is cooked or not.

The Alofa Tuvalu survey (Hemstock, 2005) accounted for the number, age and sex of people who slept in the house and the average number of people cooked for on a regular basis.

Some broad determinants of energy consumption in the household sector are shown in Table 5.1.

Field example – Tuvalu Biomass Energy Consumption Survey
Question groups used in the Tuvalu survey (Alofa Tuvalu, Hemstock, 2005):

- *Household:* size (number of males, females, under 18s sleeping in the dwelling) and the average number of people cooked for each day.
- *Kitchen appliances:* ownership (LPG stove, kerosene stove, charcoal stove, electric stove, wood-burning stove, open fire, electric rice cooker, electric kettle, other); frequency and duration of use for each appliance (boiling drinking water, cooking rice, cooking fish, etc.); number of times each appliance is used per day, per week and per month.
- *Household fuel use:* litres of kerosene used per week; bottles of gas purchased per month/year; connection to grid; amount spent on electricity annually; inclusion in solar energy programme; number of open fires per week, amount and type of wood used per fire (usually coconut husk with some shell); number of times per week charcoal is made (coconut shell charcoal). Fuel wood and charcoal was also weighed.
- *Communal cooking:* number of community events per month – open fire is usually the preferred cooking method for communal events in Tuvalu. Preparation of food for sale (via Women's Groups providing meals for schoolchildren, etc.)
- *Coconut consumption:* number consumed per household per day by humans and animals (pigs and chickens). Coconut residue available for use as bioenergy (husk and shell) can be estimated from this.

Table 5.1 *Determinants of domestic energy consumption*

Location	Climate, altitude
Social	Definition of 'household' consumption unit
	Demographic patterns, income distribution
Economics of supply	Costs in terms of price and collection effort
Cultural	Diet, fuel preferences

- *Manure production:* number of pigs owned by each household; how often pigs are cleaned out; what use the resulting slurry is put to (usually washed into the sea or used to fertilize banana and vegetable plots). This information can be used to estimate the amount of residue for use in biogas digesters.

The survey questions were delivered in person and were phrased so that certain answers could be verified by other answers (e.g. number of times the kerosene stove is used per day and the amount of kerosene purchased per week). This approach gives an indication of the validity of the data obtained. Construction materials used for domestic dwellings were not considered as previous reliable surveys detailed the amount of locally available biomass used for this purpose. In addition, biomass use for brick-making and beer brewing was not considered as this type of activity is not undertaken in Tuvalu. Handicrafts and thatching using plant fibre (pandanas leaf) was also not considered as harvesting for this purpose is not destructive and the materials are used with virtually no biomass residue production.

The results of the survey were used in the formation of a ten-year renewable energy project to implement solar, biogas and coconut oil biodiesel schemes.

When designing a survey to assess rural community bioenergy use in developing countries there are usually four broad questions to consider as detailed below.

1 *If there is an energy shortage, particularly firewood, how severe is it?* This is an important question since this assessment is the key to any bioenergy project activity or intervention. The energy situation in rural areas may be roughly gauged by a number of ranked indicators (such as localized deforestation). The accuracy of such an analysis is dependent on whether enough detailed, spatially disaggregated information on these indicators is available. For example, one indicator is fuel switching and the types of fuels being used. The absence of a charcoal industry may suggest that fuelwood is still plentiful and able to supply domestic energy needs, since charcoal only becomes economically attractive when wood becomes scarce and has to be transported over long distances.

2 *Which groups are under the greatest stress?* Once the vulnerable areas and groups of people are identified, they must be visited in the field. This is essential since indicators need to be identified accurately as some results can be misleading and therefore need to be checked. For example, fuelwood use (in terms of mass of fuelwood used) can actually be higher in areas of fuelwood shortage than it is in areas which are less stressed in terms of indigenous woodland resources. This may be because, first, in areas where fuelwood is relatively abundant consumption is lower as fuelwood is collected, stored and well dried before use. Second, in areas where fuelwood is scarce, species with poor burning qualities and wet wood may be used without drying, thus

increasing the measured weight of fuelwood required (since it is burned wet). These indicators can only be measured reliably in the field.

In the case of fuelwood and charcoal, factors such as income level and fuel availability are crucial. The switch to dung or crop residues (unless through improved technology) may signify the lack of choice for 'better' fuels and is therefore not voluntary. The use of crop residues and dung for domestic purposes may indicate fuelwood scarcity and so is often resented by the majority of its users. Strategies to increase the use of residues must acknowledge these problems and communities must be involved in any project planning process.

3   *What end uses are to be considered?* It is difficult to separate the use of fuel for different household functions such as cooking, heating and boiling water, since the fire is often used for several functions simultaneously, and the fuel store is not differentiated. It is not worth trying to disentangle the amount of fuel used for each activity, particularly in a rapid survey, as this information is not particularly useful in the design of interventions. However, it may be worth noting in a qualitative way whether there are regular fuel-using activities in addition to cooking, since this might possibly justify the future introduction of a specialized energy saving device.[2] However, for most purposes, small-scale activities (e.g. water boiling, home-consumption beer brewing, ironing, etc.) are best included under general fuel use.

Other major end uses of fuel at the household level could include cottage industries, for example, commercial beer production, and the preparation of food for sale. Outside the household there may be small industrial uses such as bakeries, brick kilns, fish smoking, and the like.

Include these activities in the demand survey if some form of intervention is possible or if total bioenergy use is required in order to assess the amount of biomass available for other purposes. Try to get details on numbers of people (women) involved, quantities of fuel used and the current costs and constraints. If, realistically, no such intervention is possible, it is best to simply note the occurrence of the activity and leave it at that.

Systematic methods of physical weighing and measuring are only possible during detailed, long-term surveys, as they require a great deal of time both during and after the survey (e.g. in assessing the moisture content of the sampled wood). This is by far the most reliable survey method. In the case of Tuvalu, the survey took place over six weeks; seasonality and climatic variations were not an issue since temperature and rainfall remain fairly constant throughout the year.

Recall by the interviewee concerning the amounts of biomass used is unfortunately not very reliable, because the concept of volumetric or weight measurement of firewood is unfamiliar to most people. The frequency of fuel gathering trips is a more reliable measure. One simple method is to ask the respondent to make up a bundle of typical size that she brings home from a trip, and then determine by questioning how many such trips are made. It is important to differentiate between seasons when estimating the number of trips made.

The bundle should then be weighed. Estimate the moisture content of the wood, as it can account for as much as half of the total weight. You can measure the moisture content of the wood in the field with a portable meter, or samples can be taken back to the laboratory. If it is not possible to make physical measurements, you can estimate the moisture content by stating whether the wood is green, partly dry or dry. This will provide a rough but useful comparison with other households or villages in the same general area.

The above refers to firewood. Measurement of the use of charcoal, crop residues and dung as fuel has always posed additional difficulties. However, it may be possible to get relative information.

Some of the factors discussed in points 1–3 above may also provide an indication of the minimum needs, as opposed to actual consumption (which may be far above or below the needs) in a specific location or socio-economic or cultural setting; the list is not exhaustive. Most of the key determinants of domestic energy use are interrelated. Moreover, they are frequently perceptions that the questioner will have to elucidate: for example, it is the perceived 'effort' of gathering fuels within the context of many other tasks, rather than actual distance or time involved that matters.

4   *How big should the sample be?* Identify the accessible population. The size of the sample depends largely on the number of sub-sectors. A very limited number of contacts or sample points are needed within each sub-group. For each group, build up a holistic picture of the prevailing demand through observation, discussion with officials and workers who are familiar with the people, and group discussions with the subjects of the demand survey. It may be possible to focus on the experience of one or two of these groups to draw out the discussion and raise comments. Consider using random cluster sampling when every member of a population belongs to a group. It is important that as many aspects as possible are taken into account when estimating supply and demand of biomass. Avoid using samples of convenience. Simple random sampling can be a desirable method of sampling under certain circumstances. It is extremely difficult, and often impossible, to evaluate the effects of a bias in sampling.

Consider the importance of getting precise results when determining sample size. In a small community, of say up to 200 people, you should try to get a response from everyone. In order for any survey sample to be statistically significant you must sample at least 5 per cent of the population, but where the population size is small the 5 per cent guideline is not accurate. For statistical analysis, accuracy *is* determined by sample size alone; however, remember that using a large sample does not compensate for a bias in sampling, so always use random sampling and multi-level analysis techniques (see 'Analysing domestic consumption' below). (The bias in the mean is the difference of the population means for respondents and non-respondents multiplied by the population non-response rate.) Make sure that all population groups within the sample area are represented.

Variability within the data generated from the questionnaires should also be assessed. This can be done by using the median as the average for ordinal data and the interquartile range as the measure of variability. Alternatively, use the mean as the average, and the standard deviation as the measure of variability. Using the mean as the average is usually the most reliable method for data generated from bioenergy surveys since the standard deviation has a special relationship to the normal curve that helps in its interpretation. Use the range very sparingly as the measure of variability. For the relationship between a nominal variable and an equal interval variable, examine differences among averages. When groups have unequal numbers of respondents, include percentages in contingency tables. For the relationship between two equal interval variables, compute a correlation coefficient.

Errors can come from:

— small sample size;
— not using random sampling;
— the use of an inadequate time-frame;
— badly designed questionnaire;
— recording and measurement errors;
— non-response problems.

IMPLEMENTING THE SURVEY

Before implementing a full-scale survey, consider asking about ten individuals to provide detailed responses from a draft of your questionnaire. This should give you some idea as to the responses you are likely to receive and will highlight any ambiguous areas. If you are planning to 'score' questions on your survey (for example: score 5 for 'strongly agree', 4 'agree', 3 'no opinion', 2 'disagree', 1 'strongly disagree'), tally the number of respondents who selected each choice, then compare the responses of high and low groups on individual items. If there is a large spread of difference in the responses it usually means that respondents found the question to be ambiguous. Testing out your questionnaire in this way will ensure that the larger-scale implementation of the survey will fulfil your research objectives by providing the type and detail of information you require in a format that will be useful.

The following points are important and should be considered for the successful implementation of a survey:

The size and composition of the survey team

A biomass consumption survey is most effectively carried out by a small, multi-disciplinary team, composed of both men and women. In most cases the team in the field should be small (at least two but not more than three people) so as not to overwhelm local facilities and the people being assessed. They should work together, as this facilitates development of a holistic view of biomass consumption through shared observations. The Tuvalu survey was carried out by two Alofa Tuvalu personnel and ten local representatives.

The number of survey teams

Placing a number of small teams in different areas is usually not very satisfactory, as this inhibits a comparison of the findings. One small team is much better able to develop the necessary sensitivity.

The length and nature of a survey visit

Initially, the team should make short, informal visits to targeted communities to interview and observe. From these 'impressionistic' surveys, the team can decide on the necessity and structure of more detailed surveys. Whatever the type of survey, it is important to obtain the opinions and perceptions of all social groups, particularly women, landless and other disadvantaged groups whose views are often neglected or poorly documented. Team members should maintain a flexible and informal approach by making regular evaluations of the progress of the survey.

The Alofa Tuvalu survey was integrated with local social organizational structures as well as accounting for grass-roots level bioenergy users via house-to-house enquiries. It is also important to explain the purpose of the survey; interviewees may then provide more relevant information or reveal aspects of fuel use that the interviewer had not considered.

Methods

Although informality is stressed, general methodologies should be worked out in the office prior to embarking on data collection. The design of the survey should direct the fieldwork so as to maximize the quantity of useful information obtained. You should attempt to understand the ongoing processes that determine consumption and supply. For example, due to Tuvalu's geographical isolation, kerosene and LPG supply was sporadic so domestic biomass energy use was higher when other fuel sources were unavailable.

The overall picture

The most effective use of resources would be to combine the consumption survey with the supply survey, as was the case with the Alofa Tuvalu survey. Any calculation of future trends should be undertaken against a background of change.

The compatibility of data

It is important to employ methodologies that allow comparisons between different groups and areas. Careful consideration as to which method to adopt must be given before the survey team reaches the field.

The distinction between assessment and estimation is important

A clear distinction must be made between assessment and estimation. Assessment is an analytical exercise which yields fairly reliable information from the respondent and that can stand up to the rigours of statistical analysis. This may be done through questionnaire survey. An estimation will deal primarily with physical processes of measurement and quantification of primary data.

Distinguish between consumption and need

There are two approaches to estimating need, rather than consumption, per household.

1   To elicit the consumer's own perceptions of the amount of fuel required to cook the basic diet and to meet space heating and lighting requirements.
2   To determine the actual energy consumption of households living at, or close to, nationally defined poverty benchmarks in different agro-climatic regions. This will provide an estimate of minimum energy needs.

As populations frequently suffer from 'disguised' shortages, with actual consumption being less than the demand need, it is important to try to establish present and future shortfalls in energy provision. For example, in Tuvalu the demand for LPG outstripped the supply.

Main factors that determine energy consumption

To obtain a complete picture of energy consumption, and to make projections for consumption under a range of different circumstances, it is necessary to understand the factors that determine consumption patterns. However, much less attention is usually paid to the determinants of consumption than to the assessment of consumption itself, although they are equally important for both the projections and assessments of proposed interventions.

Pay particular attention to biomass consumption in urban households, as domestic fuel consumption usually accounts for the greater part of biomass fuel use. But to understand energy consumption by the urban household, it is often necessary to differentiate between cities and regions as well.

The level of analysis

An initial bioenergy survey will probably be concerned with only those sectors which use most biomass. These will include villages, urban markets and those industries that use large quantities of biomass energy (e.g. bagasse in the sugar industry, fuelwood for charcoal industries) (see Table 5.2).

ANALYSING DOMESTIC CONSUMPTION

The energy consumption of the household

Household energy consumption is frequently related to household income and size. Although cooking consumes the largest proportion of energy in the domestic sector, there are many other domestic activities that require fuel. These activities may be of considerable importance in a particular locality, but may vary according to the season.

Various possible domestic end uses of fuel are summarized below.

Cooking and related
- Domestic food preparation
- Preparation of tea and beverages
- Parboiling rice

Table 5.2 *Levels of analysis of biomass consumption*

Sector	Desegregation
Urban household	Income group
Rural household	Income group
Agriculture	Large, small farms
Large-scale industry (commercial)	Food, chemicals, paper, construction, tobacco, beverages, others
Household and small-scale smithing, other	Food, brewing, pottery, industry (informal)
Transport	Air, rail, sea, water, road, private and public vehicle
Commercial	Offices, hotels, restaurants, other
Institutions	Hospitals, schools, armed services, others

- Drying food for storage
- Preparing animal foods

Other regular forms of consumption
- Boiling water for washing
- Boiling/washing clothes
- Weaving
- Drying
- Fumigation
- Space heating
- Lighting: domestic and for deterring predators
- Ironing

Occasional forms of consumption
- Food preparation and brewing for ceremonies
- Protection (warding off wild animals, repelling insects, etc.)

Determinants of energy used for cooking

Variation between households in the amount of cooking fuel used occurs for several reasons, including:

- the type of food cooked;
- the method and equipment used for cooking;
- ethnic, class or religious factors;
- the size of the household;
- income for individuals and the household.

In industrialized countries, fairly standard cooking fuels and equipment are used. However, the specific fuel consumption (SFC) in developing countries varies considerably (even when the same type of fuel is used) from about 7 to 225 MJ/kg of food.

The potential gains from using efficient cooking equipment (e.g. stoves) are enormous. A good stove can save 30–60 per cent in fuel use. Other, no less important, benefits of stoves include health and hygiene, better cooking environment, improved safety, etc.

Alofa Tuvalu results showing variation in selection of preferred cooking fuels and methods (see Tables 5.3 and 5.4).

Use of coconut husks for boiling water in Vaitupu
A total of 41 per cent of households questioned in Vaitupu use on average 1.5 kg coconut husks (air dried – the husks from six coconuts) for boiling water each day. This represents an annual use of 0.55 t biomass (or 8.8 GJ) per household

Table 5.3 *Fuel used for cooking in households in Vaitupu, Tuvalu*

Fuel/usage	Respondents:[a] %
Use gas at least once per week[b]	59
Use open fire at least once per week	94
Use kerosene every day	100
Use charcoal every day	71
Use open fire every day	59

Notes:
[a] A total of 61 per cent of respondents boiled water before drinking in Vaitupu.
[b] Gas use would be more frequent if the supply were reliable.

Table 5.4 *Fuel used for cooking in households in Funafuti, Tuvalu*

Fuel/usage	Respondents[a] (%)
Use gas at least once per week[b]	44
Use open fire at least 5 times per week	100
Use kerosene every day	100
Use charcoal at least 4 times per week	44
Use open fire every day	59

Notes:
[a] 100 per cent of respondents boiled water before drinking in Funafuti.
[b] Gas use would be more frequent if the supply were reliable.

which uses an open fire to boil water. Over the total population of Vaitupu, this represents 0.22 t per household per year (equal to the husks from 2.4 coconuts per day).

Total annual consumption of coconut husks for boiling water in Vaitupu = 55 t (887 GJ useful energy: equivalent to 21 t of oil equivalent (toe); or around 60 t of fuelwood equivalent).

Use of coconut shell charcoal for boiling water in Vaitupu

Forty-four per cent of households questioned in Vaitupu use on average 0.20 kg coconut shell charcoal for boiling water each day. This represents an annual use of 8 t (246 GJ or 6 toe) of coconut charcoal in Vaitupu. For those respondents using a charcoal stove, this represents 0.07 t charcoal per year or 1.4 kg charcoal per week for boiling water. Over the total population of Vaitupu, this represents 0.03 t charcoal use per household per year or 0.6 kg per week.

However, production of coconut charcoal has an efficiency of only 15–40 per cent, therefore 26.4 t of coconut shells with an energy value of 529GJ are required to produce the 8 t of charcoal used annually – 139,187 coconut shells are required each year to produce the charcoal. This is equivalent to 3.5 shells per day for each household with a charcoal stove. Over the total populating of Vaitupu, this represents 1.5 shells per day for boiling drinking water.

Total annual consumption of coconut shell charcoal for boiling water in Vaitupu = 8 t charcoal (246 GJ or 6 toe of useful energy). This requires 26.4 t of coconut shells to produce and 283 GJ (6.7 toe) are wasted in the conversion process.

Communal cooking in Vaitupu

One hundred per cent of respondents had at least one communal cooking activity every three months which would involve cooking on an open fire using coconut husk and fuelwood (usually cooking a pig or a large meal).

- Composition of fuel – by weight: 21 per cent fuelwood, 7 per cent coconut shell, 72 per cent coconut husk.
- Average volume of material burned = 0.5 m³.
- Material is not densely packed and mass of material = 32.3 kg (energy value = 17 GJ/t – energy value per fire = 558 MJ).
- Total weight of biomass burned = 32 t per year (543 GJ or 13 toe or 36 t fuelwood equivalent).

Use of coconut husks for boiling water in Funafuti

A total of 67 per cent of households questioned in Funafuti use coconuts for boiling water at least five times per week. On average, 1 kg of coconut husks (air dried – the husks from four coconuts) are used for boiling water each day. This represents an annual use of 0.36 t biomass (or 6 GJ) per household which uses an open fire to boil water. Over the total population of Funafuti, this represents 0.25 t per household per year (equal to the husks from 2.7 coconuts per day).

Total annual consumption of coconut husks for boiling water in Funafuti = 156 t (2500 GJ useful energy: equivalent to 60 toe).

Use of coconut shell charcoal for boiling water in Funafuti

A total of 64% of households questioned in Funafuti use charcoal stoves for boiling water on average four days per week – sometimes in conjunction with cooking. On average, 0.1 kg coconut shell charcoal is used for boiling water each day. This represents an annual use of 15 t (463 GJ or 11 toe) of coconut charcoal in Funafuti. For those respondents using a charcoal stove, this represents 0.04 t charcoal per year or 0.7 kg charcoal per week for boiling water. Over the total population of Funafuti, this represents 0.02 t charcoal use per household per year or 0.4 kg per week.

A total of 50 t of coconut shells with an energy value of 995 GJ (24 toe) are required to produce the 15 t of charcoal used annually – 261,878 coconut shells are required each year to produce the charcoal. This is equivalent to 1.8 shells per day for each household with a charcoal stove. Over the total population of Funafuti, this represents 1.1 shells per day for boiling drinking water.

Total annual consumption of coconut shell charcoal for boiling water in Funafuti = 15 t charcoal (463 GJ or 11 toe of useful energy). This requires 50 t of coconut shells to produce and 532 GJ (13 toe) are wasted in the conversion process.

ANALYSING VILLAGE CONSUMPTION PATTERNS

A complete village survey on biomass fuel consumption would include all the categories below. For a particular problem, only the relevant (and possible) categories need be selected. Multi-level statistical analysis is a useful tool for analysing many of the relationships described below.

- Pattern of energy-use for various fuel-consuming activities in different agricultural seasons, along with the methods of acquiring the fuel (ownership, collection, exchange, purchase, etc.).
- The relationship of the pattern of energy-use with family size, land holding, cattle population, income, education, urbanization, etc., so as to predict changes over time.
- Relationships among different categories of households, with regard to energy exchange or ownership of energy assets.
- Possibilities of conversion and fuel substitution under different fuel price scenarios.
- Assessment of livestock population, forest land, uncultivated wasteland, pastures, pattern of crop production, labour, etc.
- Desegregation of biomass energy sources into a number of categories, on the basis of method of acquisition, energy content, moisture content and end use.
- The views of local people (farmers, villagers, women, leaders) as to what they feel are the most important problems they face, particularly energy shortages/difficulties.
- The source of biomass supplies and time spent in collection.
- Changes in these patterns of supply within living memory and the reasons for these changes.

The collection and consumption of biomass fuels can show marked seasonal variation. Surveys should probably be repeated for the different agricultural seasons. This will provide a picture of the different patterns of energy use in relation to the major ways in which biomass is acquired – ownership, barter, free collection and purchase.

Results from the Alofa Tuvalu survey showing consumption of biomass energy for commercial activities on a community scale in Vaitupu are detailed below.

Commercial biomass energy use

Kaupule toddy production
Some 48,180 litres per year are produced from 30 trees with an average of four taps per tree. Total non-commerical toddy production on Vaitupu is 530,336 litres per year. Total toddy production in Tuvalu is given in Table 5.10.

Coconut oil production
Each of the outer islands has a Kaupule coconut oil production facility (see Figure 5.1). Vaitupu also has a second coconut oil mill which is currently not working but is in the process of being renovated.

The Kaupule mill processes around 125 coconuts per day. The price paid to coconut producers is 0.15A\$ per nut. When the mill is running, the average daily production of coconut oil is 13 litres. Two people work full time at the mill, receiving a wage of A\$48.50 each per week, and monthly electricity bills are around A\$100. Around 100 kg of coconut husk and shells are burnt each day to dry the copra. See Tables 5.5 and 5.6.

Figure 5.1 *Coconut oil mill*

Table 5.5 *Coconut oil production on Vaitupu*[a]

Total coconut oil (l/yr)	Total coconuts used per year	Average nuts per litre coconut oil	Production cost per litre (A\$/l)	Copra required (kg/l)	Pith required (kg/l)	Total annual copra production (t/yr)
3120	30,000	9.6	3.32	2.1	3.5	6.6

Note: [a] Energy value of total oil production = 105 GJ (2.5 toe).

Table 5.6 *Biomass energy required for copra production on Vaitupu*[a,b]

Total biomass burned (t/yr)	Total number of coconut husks used per year	Total number of coconut shells used per year	Energy value of biomass burned (GJ/yr)	toe
24	72,000 (18 t/yr)	31,579 (6 t/yr)	408	10

Notes:
[a] Energy value of total copra production = 185 GJ (4 toe).
[b] Value added to the process: the crushed copra (copra cake) is sold as a feed for chickens and pigs.

Fuel-using technologies

The technology employed to burn the fuel can have a considerable effect on:

- the amount of fuel required for a particular task;
- the possible complementary end uses.

It may prove important to produce a breakdown of fuel consumption by the technology employed, as was the case with the Alofa Tuvalu survey. The next step is to establish the range of fuels used.

Energy sources

A detailed breakdown of the many forms of biomass (e.g. twigs, wood, stalks, coconut husk, coconut shell, dung, etc.) is often useful, as this enables:

- separation of the needs and problems of different socio-economic groups;
- matching the projections for consumption and supply;
- identification of current and impending fuel shortages.

As the price and availability of fossil fuels are often foremost in the factors affecting biomass consumption, non-biomass energy sources should also be included in the analysis.

To obtain a complete picture of energy consumption, it is necessary to look at all other energy sources that are used in addition to biomass. These can be divided into:

1 animate
2 natural
3 secondary fuels
4 fossil fuels.

Animate   Work performed by animals and humans is the fundamental source of power in developing countries for agriculture and small-scale industry. Animals and humans provide pack and draught power, carry headloads and transport materials by bicycle, boat and cart. Draught animal power should not be seen as a backward technology, in conflict with mechanization and modernization. On the contrary, given such a contribution, and the fact that animate power is likely to be the main, feasible and appropriate form of energy for large groups of people for the foreseeable future in many countries, it should be given prime attention. Appendix 4.3 summarizes the main methods of measuring animal and human power.

Natural energy sources   Water, wind and sunlight can all provide alternative sources of energy, and are therefore potential substitutes for bioenergy via water-wheels, transport, hydroelectricity, windmills, solar devices, etc. However, natural energy sources may be only a seasonal asset in some instances.

Secondary fuels   Secondary fuels are sources of energy manufactured from basic fuels. Biogas, ethanol, charcoal and electricity all fall under this heading. Electricity is a good substitute fuel for lighting, but requires expensive technology for other uses. Continuity of supply, and servicing, and spares for equipment are often uncertain and should be considered when looking at secondary fuel use.

Fossil fuels
- Coal and coke can relieve demand for biomass fuels, particularly in industrial applications. However, distribution is often affected by dependence on imported petroleum products for vehicles.
- Petroleum products – when imported, the petroleum products kerosene, gasoline, diesel and liquid petroleum gas (LPG) (i.e. propane and butane) place a heavy demand on foreign exchange. Supply can therefore fluctuate with economic fortunes.
- Kerosene is a much favoured substitute for biomass as a domestic fuel. It has the advantage that the apparatus required for its use is fairly cheap.
- Gasoline is the main fuel for personal transport.
- Diesel is the main fuel for freight and public transport and is used for small-scale generation of electricity and other uses such as irrigation and water pumping.
- LPG meets the same demand as biogas and if it is economic and regular in supply it will have an effect on the amount of biomass used for domestic fuel. However, LPG stoves are expensive, so its use may be restricted because of this.

By combining fuel consumption data from the analysis of various end uses and fuel types, it is possible to produce a table or chart giving fuel consumption according to type and sector. Figure 5.2 provides an example of primary energy consumption in Tuvalu.

It is thus now possible to calculate the total consumption of any fuel for a particular end use. End-use (or final) consumption is, of course, quite distinct

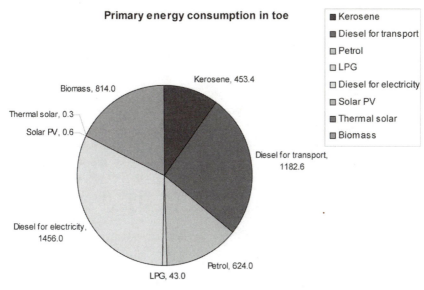

Primary energy consumption in toe

- Kerosene
- Diesel for transport
- Petrol
- LPG
- Diesel for electricity
- Solar PV
- Thermal solar
- Biomass

Kerosene, 453.4

Biomass, 814.0

Thermal solar, 0.3

Solar PV, 0.6

Diesel for transport, 1182.6

Diesel for electricity, 1456.0

Petrol, 624.0

LPG, 43.0

Source: Hemstock and Raddane, 2005

Figure 5.2 *Primary energy consumption for Tuvalu, 2005*

from total energy consumption, which includes the energy losses experienced in producing or transporting and energy form (see Figure 5.3).

Changing fuel consumption patterns

Indicators of changing fuel consumption patterns

There are five main indicators of changing fuel consumption patterns:

1   fuel collection
2   type of fuel collected
3   fuel using practices
4   marketing of fuels
5   fuel supply enhancement.

Monitoring these indicators will give a good idea of changes occurring in consumption patterns.

Fuel collection
Pointers include:

- increase in time required to collect;
- increase in distance travelled;

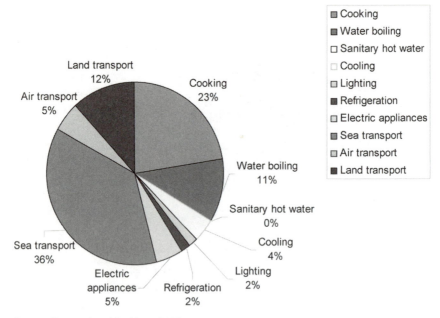

Source: Hemstock and Raddane (2005).

Figure 5.3 *Final energy consumption by end use for Tuvalu, 2005*

- change in type of collectors;
- change in transportation of fuel.

Increases in time required and distance travelled This would necessarily seem to follow from fuel scarcities, but the situation is often more complex than it first appears. While an increase in distances covered is bound to result in an increase in time spent collecting, the old and women with small children will spend more time on intensified scavenging of areas near home for leaves, roots and other low-grade fuels. Resorting to lower grades of fuel will not involve longer journeys, especially when fuel collecting is combined with other activities. In the short term, distances might even be reduced.

Change in who collects fuel In some cultures, fuel shortages may result in men joining in wood collection, particularly when there is a switch to larger parts of trees, requiring tools not normally available to women. In other cases, children may be withdrawn from school either to collect fuel themselves or to carry out other tasks to give adults more time to fetch fuel.

Change in means of transport The necessity to travel longer distances may mean that bicycles, handcarts or even draught animals are used to transport the fuel, particularly if men are now involved. This may result in less frequent journeys.

However, if it becomes necessary to collect from rough terrain this may dictate a return to headloading.

Change in type of fuel collected
The indicators to look for here are:

- change from dead to green wood;
- change to younger trees;
- change to less preferred parts of trees;
- change to less preferred species;
- change to species with other valuable products;
- change to residues.

Changing from dead wood   Switching to green wood, younger trees or less preferred parts or species of trees or plants is a progressive process that will have an increasing effect on fuel-using practices.

Changing the rotation cycle   A shortening of the length of the rotation cycle under shifting cultivation is an indicator of a growing shortage of agricultural land, but will mean a change in fuel consumption to younger, smaller growth.

Switching from fuel species   Changing from fuel species to those with other valuable products is often gender related. Women are initially reluctant to take fuelwood from trees that provide fruits and other foods, medicines and craft-making material. Men will wish to reserve those species which generate a cash income or furnish implements and tools.

Changing to agricultural and animal residues   High-grade crop residues are always popular fuels, as they are generally easy to obtain and store. However, when fuel is scarce, people may resort to low-grade residues and dung. The use of these low-grade fuels sometimes carries a social stigma.

Fuel-using practices
Changing fuel-using practices are observed through:

- an increase in duration of cooking times;
- increase in intensity of cooking;
- introduction of fuel-saving devices;
- reduction in fuel-using activities;
- change to consumption patterns using less, or more, fuel.

An increase in the duration or intensity of cooking and the introduction of fuel-saving devices   When only low quality fuel is available, there is an increase in the time required for cooking. Where household income allows, there may be a switch to stoves with a greater efficiency, rather than devoting more time to collecting fuel.

Reduction in the amount of fuel-using activities   This may involve a cutback in household industry. However, before income-generating activities are curtailed, there may be switches to foods that require less cooking, or fewer cooked meals may be eaten.

Change to consumption patterns using less, or more, fuel   The introduction of a substitute fuel such as kerosene will reduce overall consumption, but the need to generate income to pay for it might drive a lower income household into starting an enterprise such as brewing, which would raise their overall consumption of fuelwood.

Marketing of fuels
Market indicators include:

- an increase in the range of fuels bought and sold;
- an increase of marketed fuels in total consumption;
- an increase in the cost of individual fuels;
- an increase in household expenditure on fuel.

An increase in the range of fuels commercially transacted   The development of a market economy will change consumption both by the import of substitute fuels and the export of charcoal and fuelwood.

An increase in the proportion of marketed fuels in the total consumption   In some areas this will denote a reduction in the availability of biomass fuel, but in others it is a sign of increasing urbanization and prosperity, for example, this is often the case with charcoal.

An increase in the cost of individual fuels   The price of substitute fuels is so tied up with national and international economics that an increase in the price of a fuel such as kerosene does not usually indicate an increase in demand. However, an increase in the cost of kerosene will usually lead to an increase in the consumption of biomass fuels by the many households where kerosene is only marginally affordable. It may also lead to a price increase in other substitute fuels.

An increase in the proportion of household expenditure devoted to fuel   Before this is used as an indicator of the increasing use of commercialized fuels, the price trends of commercial fuels must be carefully considered.

Supply enhancement
This is recognized by:

- a change of cropping pattern;
- increase in planting of fuel crops.

Change of cropping pattern   Most changes in cropping (or alterations in the number of livestock, or the abandonment of shifting agriculture for settled occupation) will result in differing amounts of residue being available as fuel.

If crop residues constitute a major part of the existing fuel supply, then any change in cropping patterns will affect fuel availability and hence consumption. However, the significance of residues in the household economy may be reduced if new farming ventures provide an increase in income.

Increased planting of fuel crops   Fuel crops are often planted as cash crops for sale in urban areas. Therefore, the appearance of fuel crops does not necessarily imply a local change in fuel consumption patterns, although the loss of residues from crops replaced by trees may affect lower income households. On the other hand, fuelwood demand is increasingly being met from forest stocks, which are decreasing at an alarming rate in some countries.

Variations in fuel consumption

Variations between localities
There are often large variations in biomass consumption between neighbouring localities. To capture these variations, it is often desirable to select clusters of communities from several regions. The extent to which activities vary between communities will determine the number and distribution of locations covered.

Several factors can cause variations between neighbouring localities, including:

- different types of food cooked;
- different methods or means of cooking (baking instead of boiling, open fire or stove);
- different practices associated with different ethnic groups, economic classes and religions;
- different income groups. Disaggregation by income group is generally useful for urban consumers. For example, those less well off are almost exclusively dependent on biomass; middle income households use charcoal, kerosene, gas or electricity, while high income households use combinations of charcoal, kerosene, gas, electricity or wood, depending on what is being cooked and the relative prices of the different fuels.

Variation of fuel consumption over time
Variations in fuel consumption are either short term, or seasonal and annual.

Short term variation   Fuel use recorded over a number of consecutive days can sometimes show large fluctuations. There are several factors that may explain this variation.

- The amount of fuel required from day to day may vary for the following reasons:
  — Some fuel using activities are only undertaken on certain days. Brewing is not a daily occupation, and food for sale is more likely prepared for local market days. Many household industries require fuel intermittently, e.g. steeping woven cloth or firing pots.
  — The number of guests and the members of the household eating elsewhere may vary considerably from day to day. Although the overall exchange in hospitality may equalize fuel consumption over time, the effect on day-to-day consumption may be large.
  — Similarly, the number of hired employees fed (or household members employed and fed elsewhere) can vary daily. The irregular nature of household industries and the interchange of employer/employee roles between neighbours with differing enterprises will alter daily rates of consumption. Larger agricultural employers are likely to hire labour intermittently.
- Changes in the behaviour of the consumer:
  — A change of user can cause a considerable change in consumption, depending on the competence of the new user. This is particularly important when an open fire is used and the quality of the fuel is variable.
  — The amount of time the user has at his or her disposal may vary. Pressure from other commitments may lead to shortage of time for collecting or economic use of fuel, and may give rise to excessive consumption.
- Other factors for consideration:
  — Weather conditions can make important differences in consumption. When the fire is exposed, windy conditions will make the fuel burn quickly. Unseasonable temperature changes may need additional space heating, and unexpected rain storms will result in damp wood with a poor performance and difficulties in collection and storage.

These short-term variations are clearly of critical importance for the interpretation of any physical measurements of consumption that are made. But even where data are collected verbally, variations may mean that the concept of a 'normal' level of consumption is relatively meaningless.

Seasonal or annual variations mean that it is important for proper assessment of biomass fuel use over the whole year. Fuel consumption may also vary to a greater or lesser degree on a seasonal basis or from year to year. Some of the causes of this variation are listed below.

Seasonal variation

- The amount of fuel required may vary through seasonal change in fuel-consuming activities:

- — The composition of the household may vary with the seasonal demands for agricultural labour. Extra workers may be hired. Family members may get local employment, or migrate seasonally to find work. Small-scale industries such as brick making or tobacco curing are also seasonal, and may require hired labour.
- — The number of ceremonies varies throughout the year. Social or religious gatherings and ceremonies do not occur at regular intervals throughout the year and may distort seasonal fuel consumption.
- — Availability of food – more food is usually available after the harvest than before it.
- — The portions of the diet eaten raw or cooked may vary, together with the quantities of food available for consumption.
- — Temperature and the need for surface heating – at higher altitudes or latitudes extra fuel and food is consumed for space heating during the cold season.
- The availability of fuel varies with the seasons.
  - — Crops providing residues for fuel have specific harvest times. Periodic activities that require fuel are often undertaken after the harvest when crop residues are abundant, particularly if storage is difficult or labour is available.
  - — The weather affects the ease with which fuel is gathered. Except where large-scale storage is available, rainy seasons affect both the quantity of fuel it is possible to collect and its burning qualities. Long periods of windy weather also lead to extra fuel consumption.
  - — Agriculture and other labour-intensive activities affect the time available for collecting fuel. Demand for labour, particularly women's labour, is greatest at sowing and harvest time. Time available for collecting fuel and cooking is thus reduced. In the dry season, the need to fetch water from greater distances may also compete with fuel collection.
  - — Storage of fuel can be a problem. Access to storage facilities may depend on wealth and status. High income households have the space and capital to build stores and to pay labour to fill them. Those with lower incomes usually do not have large quantities of crop residues to store. Different types of storage will vary in their vulnerability to damp. Some woods do not last as long in store as others and may be more susceptible to insect attack.

Annual variations
- Annual weather variations may affect the growth of plants of all types, and therefore the amount of residue (and to a lesser extent wood) available must influence the rate of regeneration of fuelwood supplies.
- Differing trends in market prices will determine the crops grown and the residues available. Timber may be grown to satisfy urban fuel demands, but

this is likely to diminish rather than increase fuel supplies. An increase in the price of fossil fuels will transfer both urban and rural demand to fuelwood. Conversely, a decrease in the cost of fossil fuels may reduce the demand for fuelwood. However, once a charcoal supply chain is established it is likely to remain fully active.

PROJECTING SUPPLY AND DEMAND

In order to make meaningful projections of supply and demand, data must usually reflect total above-ground woody biomass present in an area that could be used as a source of energy, as well as present consumption. It should be borne in mind that woody biomass has many other applications with even higher value than fuel such as timber, plywood, pulpwood, etc. Thus, only a small proportion (e.g. branches, tops, unmerchantable species, etc.) might end up as fuelwood. The initial end result of any biomass inventory should be a database constructed from tally-sheets for each of the samples (e.g. number of trees or bush species with measurements such as stem diameter, crown diameter, etc.). Measurement of parameters can be used in combination with other data, such as tree weight tables, to determine the standing stock of woody biomass. The result should be a database that shows for each strata or vegetation type the weight of woody biomass per unit area by size classes. Sustainability of the biomass resource can then be assessed by measuring consumption against availability (see Table 5.7).

There are various methods for predicting supply and demand:

- constant-trend based projections;
- projections with adjusted demand;
- projections with increased supplies;
- projections including agricultural land;
- projections including farm trees.

From the point of view of a project planner it may be wise to use more than one method of projecting supply and demand.

Constant-trend based projections

These assume that consumption and demand grow in line with population growth and that there is no increase in supplies (Table 5.7). It is a useful way to identify any resource problems and possible actions to bring supply and demand into a sustainable balance. Essentially, consumption grows with population growth and supplies are obtained from the annual wood growth and clear felling of an initially fixed stock of trees. However, as wood resources decline, costs will increase and consumption will be reduced by fuel economics and substitution of other fuels.

Table 5.7 *Estimates of national woody biomass stocks and sustainability in Zimbabwe*

Study	St. Stock[a] (Mt)	MAI[b] (Mt)	MAI as % St. Stock	Energy content (PJ)		Carbon content (Mt)		Sustainability[c] (PJ)						Surplus (C)[d] (Mt)					
				St. Stock	MAI	St. Stock	MAI	1985a	1985b	1985c	2000a	2000b	2000c	1985a	1985b	1985c	2000a	2000b	2000c
Gov. of Zim. (1985)	312	13	4	4815	195	125	5.2	-338	-246	-49	-794	-624	-253	-9.0	-6.56	-1.3	-21.2	-16.6	-6.8
Hosier et al (1986)[e]	666	9	1.4	11,255	152	266	3.6	-381	-289	-92	-837	-667	-296	-10.6	-8.16	-2.9	-22.8	-18.2	-8.4
Millington et al (1989)	1,502	47	3.1	22,530	713	601	18.8	180	272	469	-276	-106	265	4.6	7.04	12.3	-7.6	-3.0	6.8
ETC (1990)	769	32	4.2	11,535	480	308	12.8	-53	39	236	-509	-339	32	-1.4	1.04	6.3	-13.6	-9.0	0.8
ETC (1993)	1447	35	2.4	22,705	525	579	14.0	-8	84	281	-464	-294	77	-0.2	2.24	7.5	-12.4	-7.8	2.0

Notes:

a St. Stock = Standing Stock.

b MAI = Mean Annual Increment.

c Sustainability = MAI minus Production/Harvest and is a measure of surplus as it indicates either 'excess' or 'gain' in MAI (positive value) and the energy content or the standing stock required (negative value) to fulfil demand, which can also be expressed as the amount of carbon stored (positive value) or released into the atmosphere (negative value). With the following assumptions:

- the total biomass flux (production and harvest from crops, 197 PJ, forestry, 244 PJ, and livestock, 92 PJ) based on a 533 PJ average for the years 1985–1989 (equivalent to 64 GJ per capita for the base year 1985);
- biomass flux (production and harvest from crops, 197 PJ, and forestry 244 PJ) based on a 441 PJ average for the years 1985–1989 (equivalent to 53 GJ per capita for the base year 1985);
- biomass flux (production and harvest from forestry) based on a 244 PJ average for the years 1985–1989 (equivalent to 29 GJ per capita for the base year 1985).

d Surplus = MAI (carbon content) minus annual use (carbon content) and is an indication of carbon flux (using the same assumptions as Sutainability[c]).

e Energy content for wood used in the Hosier et al (1986) study = 16.9 GJ/t.

Source: Hall and Hemstock, 1996; Hemstock and Hall, 1995.

Projections with adjusted demand

These provide a useful step to examine reductions in per capita demand and corresponding effects in declining wood resources. The adjustments can then be related to policy targets, such as improved stove programmes or fuel substitution.

Projections with increased supplies

Wood supplies can be increased by a variety of measures, such as better management of forests, better use of wastes, planting, use of alternative sources such as agricultural residues, etc. Targets for these additional supply options can easily be set by estimating the gap between projected woodfuel demand and supplies.

Projections, including agricultural land

In most developing countries the spread of arable and grazing land, together with commercial logging in some areas, is a major cause of tree loss. When land is cleared by felling and burning, the result is greater pressure on existing forest stocks for fuelwood. If the wood cleared is used for fuel this will contribute towards releasing this pressure.

Projections including farm trees

Trees have multiple uses, for example fruit, forage, timber, shelter, fuelwood, etc. Farm trees, which are fully accessible to the local consumers, are often a major source of fuel in many rural areas and hence should be included in any projection models.

CURRENT AND PROJECTED ENERGY PRODUCTION AND WASTE IN TUVALU

The Alofa Tuvalu survey results for coconut production and use in Tuvalu are detailed in the sections that follow and in Tables 5.8–5.16.

Coconuts used for human and animal consumption

Table 5.8 *Coconut pith fed to the pigs in Tuvalu*

Total number of pigs[a]	Total number of coconuts used per day for feeding pigs in Tuvalu	Number of coconuts used per year for feeding pigs	Energy content of pith fed to pigs (GJ/yr)
12,328	13,534	4,940,266	25,571 (608 toe)

Note: [a] Using Alofa Tuvalu estimate for the number of pigs in Tuvalu.

Table 5.9 *Coconuts used for human consumption in Tuvalu*

Total number of people	Total number of coconuts used per day for human consumption in Tuvalu	Number of coconuts used per year for human consumption	Energy content of pith consumed by people (GJ/yr)
9,561	2,930	1,534,094	7,941 (189 toe)

Table 5.10 *Toddy production in Tuvalu*

Total number of toddy trees in Tuvalu	Total number of toddy taps in Tuvalu	Estimated total toddy production (l/yr)
1,636	4,931	1,979,818

Commercial biomass energy use

All the outer islands have a small coconut oil production facility like the one detailed for Vaitupu.

Table 5.11 *Coconut oil production in Tuvalu* [a]

Total coconut oil (l/yr)	Total coconuts used per year	Average nuts per litre coconut oil	Production cost per litre (A$/l)	Copra required (kg/l)	Pith required (kg/l)	Total annual copra production (t/yr)
16,800	198,000	10	3.92	2.2	3.6	43.56

Note: [a] Energy value of total oil production = 671 GJ (16 toe).

Table 5.12 *Biomass energy required for copra production in Tuvalu* [a,b]

Total biomass burned (t/yr)	Total number of coconut husks used per year	Total number of coconut shells used per year	Energy value of biomass burned (GJ/yr)	toe
192	576,000	252,632	3,264	78

Notes:
[a] Energy value of total copra production = 1220 GJ (29 toe).
[b] Value added to the process: the crushed copra (copra cake) is sold as a feed for chickens and pigs.

Wasted energy and total biomass energy use

- The total number of coconuts used (to feed to pigs and people) is 6,672,361 (energy available from husk and shells 52,044 GJ; 1,239 toe).
- The total number of coconut husks used as a fuel (domestic and commercial) = 5,136,242 (with an energy value of 20,545 GJ or 498 toe).
- The total number of coconut shells used as a fuel (domestic and commercial) = 2,965,101 (with an energy value of 11,267 GJ or 268 toe).
- Total biomass energy consumption (domestic) = 31,350 GJ (746 toe) (useful energy after some conversion to charcoal = 26,112 GJ or 622 toe).
- Total biomass energy consumption (commercial) = 2,856 GJ (68 toe).

Table 5.13 *Wasted energy from coconuts*

Total number of shells unused each year	Energy value of unused shells (GJ/yr)	Energy value of shells (toe)	Total number of husks unused each year	Energy value of unused husks (GJ/yr)	Energy value of husks (toe)	Total energy available from unused husks and shells[a] (GJ/yr)
3,707,260	14,088	335	1,536,118	6,144	146	20,232 (482 toe)

Note: [a] The unused husks and shells (which total 482 toe) would be ideal fuel for a series of gasifiers for electricity generation for each island's respective grid.

Table 5.14 *Wasted energy from pigs*

Total number of pigs in Tuvalu (head)	Total annual production of pig manure (t/yr)	Total annual production of energy from pig manure (GJ/yr)	Total amount of manure available annually for use in biogas digester[a] (t/yr)	Total amount of energy available annually for use in biogas digester[a] (GJ/yr)
12,328 (60% = 7,397)	3,600	32,400 (771 toe)	2,106	19,440 (463 toe)

Note: [a] Assumes 60% collection efficiency as some waste is used for compost and some will be difficult to collect.

Estimate of coconut production in Tuvalu

Table 5.15 *Coconut production in Tuvalu*

Total number of hectares under coconut	Estimate of total number of productive trees	Estimate of total number of trees over 60 years	Total coconut production (nuts/year)	Number of hectares requiring replanting	Theoretical availability of coconuts for biomass energy production[a] (nuts/year)
1,524	267,760	68,929	14,141,100	391	7,468,739

Note: [a] Total production minus use.
Source: Based on Dept. Lands and Survey, 2004 – data from 1986–88; Trewren (1984) and Seluka et al (1998); Alofa Tuvalu survey, 2005 (Hemstock, 2005).

Table 5.16 *Theoretical production of coconut oil from unused coconuts in Tuvalu*

Total number of litres of oil per year	Energy value of coconut oil (GJ/yr)	toe of oil produced	Copra production[a] (t/yr)	toe of the copra produced	% of boat fuel replaced by coconut oil
777,184	26,168	623	1,643	39	56

Note: [a] This estimate is based on coconut production per productive tree in Tuvalu and is more conservative than the figure for copra production of 1.2 t/ha used by Trewren (1984).[3] However, it is less than Trewren predicted if you use his figures of 215 trees/ha; 60 nuts/tree and 0.187 kg copra/nut.

CONCLUSIONS

As stated in the Introduction to this chapter, two aims of the Alofa Tuvalu were to assess the amount of biomass energy resource available for biogas digestion and coconut oil biodiesel production projects, and to ensure that they were sustainable in terms of current biomass production.

From the tables it appears that coconut production in Tuvalu is sufficient to meet requirements for domestic biomass energy, current indigenous food use and 56 per cent of current fuel use of the outer island boats (see Figures 5.2 and 5.3). The boats use half the total fuel oil imported into Tuvalu. Therefore, current coconut productivity in Tuvalu could easily replace 25 per cent of current total fuel oil use. Payment to copra growers would total A$575,093 annually at prices currently agreed with producers on Vaitupu.

As emphasized previously, biomass energy comprises many components, for example, production, conversion, use, conservation, etc., each further subdivided into many equally important sub-components. Accurate consumption and productivity assessments are a fundamental tool for project planning and subsequent management of biomass resource development and conservation programmes. Without reliable data for both consumption and availability it is difficult to undertake meaningful policy and planning decisions. Many decisions in the past have been made from a very poor footing which has resulted in many project failures.

Notes

1 French-based international NGO, Alofa Tuvalu, 30 rue Philippe Hecht, 75019 Paris, France.
2 Cooking and boiling water were assessed separately in the Alofa Tuvalu survey and as a result the introduction of solar water heaters for providing sanitary and drinking water has been recommended.
3 Trewren, K. (1984) *Coconut Development in Tuvalu*, Ministry of Commerce and Natural Resources, Funafuti, Tuvalu.

References and Further Reading

Dept. of Lands & Survey (Tuvalu) (2004) Geographical information showing the land areas of the islands of Tuvalu.

ETC (1990) *Biomass Assessment in Africa*, ETC (UK) Ltd in collaboration with Newcastle Polytechnic and Reading University, World Bank, Washington, DC, USA

ETC (1993) *Estimating Woody Biomass in Sub-Saharan Africa*, ETC (UK) Ltd in collaboration with Newcastle Polytechnic and Reading University, World Bank, Washington, DC, USA

Government of Zimbabwe (1985) 'Geographical extent of vegetation types, estimates of total and accessible areas surviving in 1984 and their growing stock increment', in Millington et al (1989) *Biomass Assessment – Woody Biomass in the SADCC Region*, Earthscan Publications, London, UK

Hall, D. O. and Hemstock, S. L. (1996) 'Biomass energy flows in Kenya and Zimbabwe: Indicators of CO_2 mitigation strategies', *The Environmental Professional*, 18, 69–79

Hemstock, S. L. and Hall, D. O. (1995) 'Biomass Energy Flows in Zimbabwe', *Biomass and Bioenergy*, 8, 151–173

Hemstock, S. L. (2005) *Biomass Energy Potential in Tuvalu*, (Alofa Tuvalu), Government of Tuvalu Report

Hemstock, S. L. and Raddane, P. (2005) *Tuvalu Renewable Energy Study: Current Energy Use and Potential for Renewables*, Alofa Tuvalu, French Agency for Environment and Energy Management – ADEME, Government of Tuvalu

Hosier, R. H., Katarere, Y., Munasirei, D. K., Nkomo, J. C., Ram, B. J. and Robinson, P. B. (1986). *Zimbabwe: Energy planning for national development*, Beijer Institute, Stockholm, Sweden

Millington, A., Townsend, J., Kennedy, P., Saull, R., Prince, S. and Madams, R. (1989) *Biomass Assessment – Woody Biomass in the SADCC Region*, Earthscan Publications, London, UK

Seluka, S., Panapa, T., Maluofenua, S., Samisoni, Tebano, T. (1998) *A Preliminary Listing of Tuvalu Plants, Fishes, Birds and Insects*, The Atoll Research Programme, University of the South Pacific, Tarawa, Kiribati

Trewren, K. (1984) *Coconut Development in Tuvalu*. Ministry of Commerce & National Resources, Funafuti, Tuvalu

Statistics bibliography

http://obelia.jde.aca.mmu.ac (statistics and survey design information)
www2.chass.ncsu.edu/garson/pa765/statnote.htm (online statistics text book)
http://duke.usask.ca/~rbaker/stats.html (basic statistics)
www.pp.rhul.ac.uk/~cowan/stat_course_01.html (more statistics)

Bethel, J. (1989) 'Sample allocation in multivariate surveys', *Survey Methodology*, 15, 47–57

Braithwaite, V. (1994) 'Beyond Rokeach's equality-freedom model: Two dimensional values in a one dimensional world', *Journal of Social Issues*, 50, 67–94

Feldt, L. and Brennan, R.(1989) 'Reliability' in *Educational Measurement*, Linn, R. (ed.), Macmillan Publishing Company, 105–146

Gibbins, K. and Walker, I. (1993) 'Multiple interpretations of the Rokeach value survey', *Journal of Social Psychology*, 133, 797–805

Goldstein, H. (1995) *Multilevel Statistical Models*, Halstead Press, New York

Longford, N. (1993) *Random Coefficient Models*, Clarendon Press, Oxford

Valliant, R. and Gentle, J. (1997) 'An application of mathematical programming to a sample allocation problem', *Computational Statistics and Data Analysis*, 25, 337–360

6

Remote Sensing Techniques for Biomass Production and Carbon Sequestration Projects

Subhashree Das and N. H. Ravindaranth[1]

INTRODUCTION

The focus of this chapter is on the estimation of woody biomass production for use as a source of energy. The sources of woody biomass for energy include forests, tree plantations and agro-forests. Biomass can be harvested in a sustainable manner from primary (or secondary) forests or produced sustainably in plantations.

Forest managers, plantation companies, and bioenergy utility managers need to estimate or project biomass production and growth rates from forests and plantations for planning bioenergy projects. The evaluation of biomass production or growth rate requires an estimation of area under forests or plantations at a given time, or area changes over a period of time, and estimates of the weight of trees or components of trees (tree trunk, branches and leaves) in a unit area. Traditionally, forest mensuration techniques are used for estimating forest biomass production. The other methods include using allometric equations and tree harvesting. These methods require elaborate periodic field measurements and are often not cost effective.

Satellite imagery based remote sensing techniques provide an alternative to traditional methods for estimating or monitoring or verifying an area of forest or plantation, and biomass production or growth rates. Remote sensing techniques provide spatially explicit information and enable repeated monitoring, even in remote locations, in a cost-effective way. These techniques have been routinely used for monitoring area under forests, plantations, field crops, settlements and infrastructural facilities, particularly at regional and national scales. Remote sensing techniques for estimating growing stock of biomass or its productivity are, although feasible, yet to become popular.

This chapter presents bioenergy utility and forest or plantation managers with various remote sensing techniques for estimating, monitoring or verifying biomass production or growth rates. Further, these techniques can be used for estimating carbon stock changes in carbon sequestration projects. The focus is on assessing the utility of the techniques for estimating biomass at a project or large plantation or at a landscape level, and not for national or global forest biomass assessment.

REMOTE SENSING AND BIOMASS PRODUCTION

Remote sensing can be defined as the art and science of obtaining information about an object without being in direct physical contact with the object. The possibilities for measuring important biophysical characteristics through this technique are immense. Over time it has emerged as one of the premier scientific tools that may be used as an analytical aid towards better management of the world's forest resources. Repeated and consistent monitoring of large areas of forests is possible only with the use of remotely sensed techniques. In the case of forestry, remote sensing can be useful in the identification and analysis of forest or plantation areas, i.e. their location and size, state of degradation and the level of human pressure visible through deforestation, fires and agroforestry. With high-resolution satellites, certain physiognomic parameters related to various cover classes permit the discrimination of forest, woodland and shrubland, while floristic parameters permit the determination of broadleaved, coniferous and mixed stands. Satellite remote sensing can also assist in forest management by providing information on accessibility, for example, topography, paths or roads, and also permit yearly or even monthly monitoring of main forest stands and logging over very large areas such as provinces or countries.

Automated image analysis techniques make interpretation and estimation of forested tracts simpler. The applicability of the technique has also gained ground due to the relatively shorter time period in which the required information can be gathered, as compared to other techniques for estimating biomass. Over time, tremendous progress has been made in information generation through application of remote sensing and its subsequent integration with a geographic information system (GIS) programme. With the advancement in the development and deployment of sensor technology, there is a wide range of data types available for identification, classification, biomass estimation and measurements of forest cover types which include optical multispectral scanner, radar imagery, lidar, hyperspectral imagery, etc. Image data processing along with information extraction techniques have been extensively studied and novel methods, such as multi-sensor data fusion, have been developed for better information gathering and spatial analysis. Parallel to this has developed an understanding of how and why remotely sensed data and methods are important in forestry and forest science.

Remote sensing of forests or plantations begins with a well-designed data collection survey followed by a data preparation activity. This data may then be used to answer well-defined questions with the help of detailed spatial analysis. Critical issues of data quality, resolution, repeatability and validation also need to be considered while attempting to determine forest structure, productivity, stand growth, density mapping, etc.

The core of the remote sensing techniques lies in understanding the relationship between forest stand parameters and spectral responses, depending on the characteristics of the study area. A good understanding of the subject is a prerequisite for using image bands for modelling estimates of biomass. Most research focuses on developing relationships between structural parameters of forests or tree plantations such as basal area, biomass, crown closure, tree height, diameter at breast height (DBH) and spectral responses. Hence, a comparative analysis of spectral responses and biophysical parameters in different forest stand structures and environmental conditions is useful while attempting to evaluate forests for biomass production.

The broad steps involved in using remote sensing techniques for biomass or carbon sequestration projects are given in Figure 6.1.

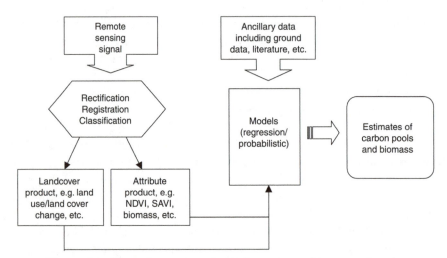

Figure 6.1 *Flow of remote sensing information to biomass estimation*

Remote sensing techniques

Introduction

Based on the different kinds of sensors available, remote sensing data may be acquired at varying spatial scales as well as varying temporal scales. The sensors may be broadly classified as:

- photographic (air);
- scanning (air and space);
- RADAR (radio detection and ranging);
- LIDAR (light detection and ranging);
- videography.

Aerial photography

Traditionally, forest management questions were answered with field studies that were very detailed and accurate but elaborate in terms of time and cost. Also, such studies were devoid of accurate geographic coordinates and hence did not allow tracking of exact forest stand locations. The advent of aerial photography catered to the basic requirements of locational capabilities, and is one of the most widely used forms of remote sensing of forest cover. It has found applications in forest assessment, inventory, monitoring and classification. Cameras used for aerial photography are the simplest and the oldest forms of aerial sensors used for remote sensing of the earth's surface features. Cameras are framing systems and are considered passive optical sensors that use a lens or a system of lenses to capture ground detail. Cameras can be of different types, namely, single-lens mapping, multiple-lens mapping, panoramic and digital. The spectral resolution of camera lenses is usually coarse; hence, aerial photography is more useful when fine spatial data is more important than spectral information.

There are various kinds of photographic films available with respect to the types of emulsions used. Photographic films are sensitive to light from 0.3 μm to 0.9 μm in wavelength covering the ultraviolet (UV), visible, and near infrared (NIR):

- black-and-white film (panchromatic);
- black-and-white infrared film;
- normal colour film;
- colour-infrared film.

Panchromatic films are sensitive to the UV and visible portions of the spectrum and produce black-and-white images. They are the most common type of films used. UV photography also uses panchromatic film but with a filter that blocks out the visible portions of the spectrum. The use of panchromatic films is restricted by atmospheric scattering and absorption, due to problems of atmospheric scattering and absorption. Black-and-white infrared films are sensitive to the entire 0.3–0.9 µm wavelength range and are useful for detecting differences in vegetation cover, due to its sensitivity to IR reflectance. The scale of the photograph is also an important consideration when aiming at forestry applications and ranges from 1:70,000 to 1:5000 depending upon the objective of the study and terrain conditions. The commonly used platforms for cameras are helicopters, aircraft and spacecraft. Aerial photography can be used to cover areas ranging from a few hundred to a few thousand hectares of forests, plantations and agro-forestry systems.

Tree parameters, DBH, height and tree crown calculations

Diameter of the trees is an important indicator of size or volume and is generally measured at 1.30 m height from the ground (see Chapter 3). Such measurements are not possible directly from aerial photographs, since they require very large-scale oblique photographs. Most DBH measurements are made using a combination of field studies and photographs. Crown diameters are estimated from the photographs and subsequently used in regression equations to estimate the DBH. Stereoscopic pairs of photographs are needed to measure the crown diameters with the help of instruments such as the crown diameter gauge or the dot grid. This technique has been widely used in the measurements of DBH in temperate as well as tropical forests. For measuring the crown diameter, the preferred scale of photography is 1:5000. Some researchers have suggested the measurement of crown area instead of crown diameter to account for the irregularity in the shapes of crowns of different species. In some cases crown area has been found to be more closely correlated to DBH than crown diameter. Another mehod commonly used in regression models is mean tree height. Sometimes it may be difficult to estimate tree heights, due to closed tree canopies or unknown differences in the elevation of the terrain. In order to minimize such errors, the Criterion Model 400 LASER rangefinder or other similar devices may be used, which report accuracies of ±9 cm to estimate tree height in the field, and which can be later used as a surrogate measure. Based on the type of forest, particular regression models can then be applied to obtain DBH or height values.

Volume estimations

Volume may be expressed in different ways, depending on whether it is being estimated for a single tree, a forest stand or a plantation, and as gross or net volume, or total or merchantable volume, etc. Wood volume is perhaps one of the most important parameters assessed. It is difficult to directly estimate the tree and stand volume and may prove very costly. Therefore it is preferable to carry out volume estimations of a forest stand using regression equations, which have a set of dependent and independent variables.

There are many categories of usable volume equations, one of which is employed with large scale photography (LSP), and is known as the aerial tree volume equation. In such equations, one of the independent variables is measurable from the aerial photographs. Such equations are usually applied for a group of species rather than for individual species.

Advantages and disadvantages of the use of aerial photographs

The major advantages of using aerial photography over traditional ground-based mapping systems for biomass estimation are as follows:

- covers a large area of the land at approximately the same scale;
- high resolution;
- better interpretation of features with stereoscopic vision;
- can be used for places which are ordinarily inaccessible;
- easy to make copies and store;
- easier availability of conventional photographs;
- measurements are possible if the scale is known;
- established and accepted methodology.

However, there are some disadvantages associated with aerial systems, namely:

- absence of georeferencing makes digitization difficult;
- may have tilts and errors, such as relief displacement, which may give incorrect measurements from the photograph;
- positional location and scale are approximate;
- ground features may be obscured by other features;
- lack of contrast in colours;
- cost can prove to be high for small-scale projects;
- relatively long time required to obtain final prints.

Multispectral scanning systems

Optical remote sensing makes use of visible, near infrared and short wave infrared sensors to form images of the earth's surface by detecting the solar radiation reflected from targets on the ground. Different materials reflect and absorb differently at different wavelengths. Thus, the targets can be differentiated by their spectral reflectance signatures in the remotely sensed images. Vegetation has a unique spectral signature that enables it to be distinguished readily from other types of land cover.

Multispectral remote sensing is defined as the collection of reflected, emitted or backscattered energy from an object or area of interest in multiple bands (regions) of the electromagnetic spectrum. Satellite-based monitoring systems may be subdivided into low, moderate and high-resolution data. Optical sensors are characterized by spectral, radiometric and geometric performance. Satellites are usually classified according to their spatial resolution into environmental satellites (Meteosat, GOES, NOAA), medium-resolution satellites (Landsat MSS, IRS1, JERS1); and high resolution satellites (Landsat TM, SPOT, ERS-1, IKONOS, etc.). Environmental satellites are best suited to frequent (daily or weekly) monitoring of relatively large areas, such as continents, subregions or countries. They are primarily used in meteorology and oceanography and, more recently, for monitoring vegetation conditions, principally of large pastures or forested areas, at scales from 1:10 million to 1:20 million. Medium-resolution satellites are primarily represented by Landsat, with its multispectral scanner (MSS) which has been in operation since 1972. They provide imagery at small to medium scale (1:1,000,000 to 1:200,000) for land use studies, and have particular application to forestry. High-resolution satellites are more recent and have principally been used since the mid-1980s. They allow mapping at scales of up to 1:25,000 (e.g. SPOT).

Above-ground biomass has been estimated using Landsat TM, NOAA AVHRR data, etc. Most studies involve deriving a relationship between a commonly used vegetation index such as NDVI (normalized difference vegetation index) with biomass or some biophysical parameter. A large number of protocols can be used to develop biomass estimates. Below, we define a broad overview of the steps involved in the estimation of biomass for a small area on a local scale:

1   *Designing a field study:* involves designing a sampling strategy, selection of plots (size and number), collection of forest inventory data (species), measurement of variables like tree DBH, height, etc.

2   *Estimating above-ground biomass and basal area:* may be achieved by using traditional equations or regression models.

3   *Acquisition of satellite data and pre-processing:* procurement of appropriate satellite data for the same time period as the field studies and application of geometrical, radiometric and atmospheric correction. Cloud screening and sensor calibrations may also be considered.

4   *Calculation of vegetation indices:* a vegetation index is a dimensionless, radiation-based measurement computed from the spectral combination of remotely sensed data. An index is useful due to its empirical or theoretical relationship with biophysical variables. Simple, normalized and complex vegetation indices may be calculated from the satellite imagery. Table 6.1 gives a list of the commonly used indices. Image transforms (e.g. KT1, PC1-A, Albedo, etc.) may also be calculated.

5   *Integration of satellite data and spectral responses:* overlay of sample sites on

Table 6.1 *Vegetation indices used in biomass estimation*

Vegetation index	Formula
Simple ratio	
TM 4/3	TM4 / TM3
TM 5/3	TM5 / TM3
TM 5/4	TM5 / TM4
TM 5/7	TM5 / TM7
SAVI	[(NIR − RED) / (NIR + RED + L)] (1 + L)
Normalized indices	
NDVI	(TM4 − TM3) / (TM4 + TM3)
ND 53	(TM5 − TM3) / (TM5 + TM3)
ND 54	(TM5 − TM4) / (TM5 + TM4)
ND 57	(TM5 − TM7) / (TM5 + TM7)
ND 32	(TM3 − TM2) / (TM3 + TM2)
Image transforms	
KT3	0.151TM1 + 0.197TM2 + 0.328TM3 + 0.341TM4 − 0.711TM5 − 0.457TM7
VIS123	TM1 + TM2 + TM3
MID57	TM5 + TM7
Albedo	TM1 + TM2 + TM3 + TM4 + TM5 + TM7

imagery and extraction of corresponding image information for each sample site.

6   *Derivation of relation between stand parameter and vegetation index/reflectance:* calculation of correlation coefficients between selected stand parameter or net primary production (NPP), etc. and vegetation index or reflectance values. A predictive relation (e.g. biomass–NDVI relation) can be arrived at using the regression equation developed.

7   *Estimation of biomass through modelling in GIS or application of equation:* biomass can be estimated using the relationship developed in the above step or by coupling with other factors (e.g. meteorology) and subsequent modelling in GIS. Models of NPP and carbon densities can be generated using overlay techniques in GIS.

8   *Biomass estimate testing and model validation:* the calculated biomass estimates may be compared with inventory data to check for accuracy estimation. Similarly, GIS-based models may also be validated. Inter-transferability can be attempted across different zones and forest types.

An example of biomass and NPP estimation in India is that of the techniques developed by Roy and Ravan (1996). Biomass studies were carried out in deciduous natural forests. Two approaches were used to estimate biomass. The first approach employed the statistical sampling technique (SST), which involves deriving homogenous vegetation strata (HVS) using satellite data, physiography, density, basal area, etc. In the second method called 'spectral response modelling', simple linear and multiple regression models have been developed between spectral response and per unit biomass values. Wood volume and tree biomass have also been estimated in Sweden with the help of Landsat TM data (Fazakas et al, 1999). The estimation method is known as the '*k* nearest-neighbour method', in which estimates are produced for each forest land pixel in the image. However, studies suggest that such methods are applicable for rather small areas only. Foody et al (2003) tested the transferability of the tropical forest biomass relationships with biophysical parameters across Brazil, Malaysia and Thailand and neural network based approaches were found to be better correlated with biomass estimates from field surveys. Another approach is to combine Landsat TM data for estimating the age of a stand and combining it with lidar estimates for stand height and above-ground biomass. The combined data can give estimates of Net Primary Production (NPP) and net ecosystem production (NEP). GIS-based biomass modelling has been attempted by Brown et al (1993) to generate maps of Asia and Africa for the potential as well as actual biomass density.

Microwave remote sensing

Microwave systems utilize the microwave portion of the electromagnetic spectrum. Microwave sensors differ from optical scanning systems in that they can provide information on canopy structure, plant water content and soil conditions.

Radar is an acronym for 'radio detection and ranging'. It is an active system as it emits radio waves and illuminates the surface of the earth and records the energy backscattered from the terrain. 'Side-looking airborne radar' (SLAR), developed during the late 1950s can obtain images over vast regions to the left or right of the aircraft. There are two types of SLAR currently being used currently, 'real aperture radar (RAR)' and the 'synthetic aperture radar (SAR)' based on whether the antenna being used is of fixed or variable length, respectively.

The major advantages of a radar system are that the longer radio waves can penetrate vegetation canopy more deeply than the optical wavelengths. The most commonly used wavelengths in imaging radar are K (1.19–1.67 cm), C (3.9–7.5 cm), S (7.5–15.0 cm), L (23.5, 24.0, 25.0 cm) and P (30.0–100 cm). These longer wavelengths can also penetrate cloud cover, which is a hindrance to optical systems. Also, microwave systems are better able to differentiate between forest cover types and densities than visible wavelengths. Active microwave sensors (e.g. synthetic aperture radar, SAR) have a good spatial resolution, but their radiometric resolution is moderate, because of the coherent measurement method. Unlike the active sensors, passive sensors (e.g. radiometers) have good radiometric accuracy and a wide coverage but the spatial resolution is moderate (tens of kilometres). Therefore, radiometers could be used for global monitoring while sensors such as SAR provide more detailed local information (Kurvonen et al, 2002).

The various radar satellite systems which have been most commonly used for biomass estimations across the world are given in Table 6.2.

Radar backscatter increases approximately linearly with biomass until it saturates at a level that depends on the radar frequency. HV (horizontal–vertical) polarization in longer wavelengths (L or P band) is most sensitive to biomass (Sader, 1987; Le Toan et al, 1992, 1997a) because it originates mainly from the canopy volume scattering (Wang et al, 1995) and trunk scattering (Le Toan et al, 1992), and is less affected by the ground surface (Ranson and Sun, 1994). Estimation of stem volume, biomass, DBH, basal area and other forest stand parameters are possible by developing regression models between stand parameters and radar backscatter.

Reflection from soil and ground may sometimes be recorded as part of the backscatter. Nevertheless, this constraint can be overcome by using appropriately transformed models. There are different kinds of backscatter models, namely, radiative transfer models, regression models and conceptual models. Experiments have been carried out in tropical rainforests and boreal forest using synthetic aperture data (SAR) and radiometers (passive systems). Kurvonen et al (2002) have developed an adaptive inversion method to estimate the stem volume of a forest stand from JERS-1 and ERS-1 SAR images based on a semi-empirical backscattering model for the boreal forests of Finland. Radiometers were also used to monitor boreal forests using a mixed pixel classification approach. Using an adaptive inversion method, a pixel-wise stem volume was estimated using the special sensor microwave imager (SSM/I). The results suggested that satellite-borne microwave radiometry data has the potential for large-scale biomass estimations. Santos et al (2003) have studied the relationship of P-band SAR data with biomass values of primary forest and secondary succession of the Brazilian tropical rainforest and proved conclusively that P-band data could substantially contribute towards the development of models to monitor the biomass dynamics of tropical forests. Pulliainen et al (2003) have suggested an operative stem volume approach for forest stem volume estimation from Interoferometric SAR (INSAR) data based on a developed inversion technique.

Table 6.2 *Radar satellite systems for forest assessment*

Satellite	Ground resolution (m)	Spectral bands	Repetivity
Seasat (USA)	20 and 50	L-band	NA
SIR-A (USA)	40	L-band	NA
SIR-B (USA)	30	L-band	NA
ERS-1	30	C-band	3–143 days
JERS-1 (Japan)	18	L-band	44 days
SIR-C	50	X, C, L bands	NA
Radarsat (Canada)	10–100	C-band	NA

Light detection and ranging (lidar) systems

'Laser', an acronym for 'light amplification by stimulated emission of radiation', is another type of microwave sensing. The laser sensor system used for remote sensing is called 'lidar' (light detection and ranging). Lidar is an active system offering tremendous potential for monitoring forest biomass. The lidar system measures the time taken for a pulse of laser to travel between the sensor and the target. The pulse travels through the atmosphere and interacts with objects on the earth's surface, and is then reflected back to the sensor. The time for the traverse is calculated, bearing in mind that the speed of light is constant, and the intensity of the reflected radiation is measured. Remote sensing using lidar is currently carried out using small aircraft or helicopters operating at low altitudes.

The two most widely used lidar systems are the laser profiler or altimeter and the reconnaissance or mapping lidar. The Lidar-In-Space-Technology Experiment (LITE) mission was realized in 1994, the data from which gave a global view of various environmental phenomena such as biomass burning, aerosol concentration, cloud structure, etc. The major advantage of the use of lidar estimates in forestry applications is the acquisition of three-dimensional data of the forest structure, and data on canopy cover characteristics, leaf area index, crown cover and volume, etc. The ability of the laser altimeters to penetrate forest canopies through to the ground level is a further benefit.

Lidar-based forestry studies have not been conducted in large numbers across the world, and not much research has been directed towards the appropriate use of the technology. The majority of the studies are restricted to USA, Sweden, Norway, Canada and Germany. Tree height for homogeneous forest stands has been estimated using the topographic profiles obtained from altimetry. Crown cover density has been obtained using data on missing pulses, amplitude ration, number of peaks, ground area, canopy area, height, etc. Nelson et al (1988) have given a procedure to predict tree volume and tree biomass. Stem volume has been estimated at plot level using percentiles of the laser canopy height distributions combined with the ratio between the number of vegetation measurements and the total number of laser measurements. The variables used were the laser height variable, laser canopy density variable and ground variables. Tickle et al (1998) integrated a laser profiling device, a differential global positioning system and a digital video to compare tree height, projective foliage cover, crown cover and large-scale video estimates of stocking and growth stage of species with the ground surveys of the same variables in Australia. The study demonstrated the capabilities of the integration of these technologies for the generation of ground quality information. Another example of data fusion is a study by Lefsky et al (2005) where above-ground net primary production of wood has been calculated by combining lidar estimates of above-ground biomass with optically derived stand age from Landsat TM data.

High-resolution sensor systems

There are two major sources of high spatial resolution systems: digital frame cameras and high-resolution satellites. In digital frame cameras, the conventional film is replaced by a solid-state array of imaging elements. Their use allows real time viewing of results and does away with charges for printing, scanning, film processing, etc.

High-resolution satellites have now become available for commercial use. Such data has the potential to improve the accuracy of traditional forest inventory methods. The structural diversity of the forests can also be retrieved from such imagery. A major potential application for high-resolution data is to predict forest basal area and biomass from tree crown size (Read et al, 2003). Table 6.3 shows a list of the commercially available satellites with their resolutions.

Analysis of high-resolution imagery, however, remains a complicated task and requires new algorithms for image interpretation. Traditional per-pixel classifiers that were based on pattern recognition alone may not suffice in all cases for such high-resolution data. Mauro (2003) used QuickBird images to study mountainous forests of northern Italy. The study developed a methodology to estimate woody biomass using NDVI and compared results with biomass weight data. The study found that both these techniques were comparable, with a demonstrable relationship between wood biomass and NDVI, although a larger number of samples is required to confirm this relationship.

Many individual trees are distinct using high-resolution imagery like those obtained from IKONOS, indicating that it is feasible to conduct demographic studies of tropical rain forest canopy trees based on repeated satellite observations. Linking these remotely sensed data to ground data requires improved GPS positions, because it is currently difficult to obtain accurate GPS readings in tropical rain forest understoreys. Another option is to merge IKONOS

Table 6.3 *Overview of high-resolution satellites*

Satellite	Type of sensor	Resolution
CARTOSAT-1 (IRS-P5)	Panchromatic	2.5 m
RESOURCESAT (IRS P-6)	LISS IV	5.8 m
EROS A1	Panchromatic	1.8 m
IKONOS	Panchromatic	1 m
	Multispectral	4 m
IRS-1C	Panchromatic	5.8 m
IRS-1D	Panchromatic	5.8 m
QuickBird 2 (renamed as QuickBird)	Panchromatic	0.61 m
	Multispectral	2.44 m
SPIN-2	Panchromatic	2 m
	Panchromatic	1 m
OrbView 3	Multispectral	4 m

panchromatic 1 m data with 4 m resolution multispectral data to identify individual trees and estimate logging parameters. Assessments of logging parameters, forest gap formation and recovery rates over large areas are also potential areas where such imagery can be used.

The protocol followed by Read et al (2003) for the prediction of basal area and biomass from IKONOS imagery is as follows:

1   Acquisition of IKONOS image and merging of 1 m panchromatic and 4 m multispectral data to derive 1 m pan-sharpened image.
2   Image georeferencing using ground control points based on tree crowns identifiable on the image.
3   Measurement of the crown area index and trunk diameter for each tree on the ground.
4   Determination of the position of the vertical projection of the canopy edge on the forest floor with a clinometer.
5   Calculation of a ground-based index of crown area and correlation with crown area digitized on the image.

The above study found that there are new avenues in using high-resolution data for scales of 10–1000 m² areas. The crown area digitized from the imagery and the ground values were found to be significantly correlated and therefore crown area estimates can be used to predict canopy tree diameters and hence a significant portion of biomass.

Hyperspectral imagery

Imaging spectroscopy is defined as the simultaneous acquisition of images in many relatively narrow, contiguous and/or non-contiguous spectral bands throughout the ultraviolet, visible and infrared portions of the spectrum. Imaging spectroscopy allows simultaneous acquisition of data in hundreds of spectral bands, facilitating greatly detailed study of the earth's resources.

The NASA EO-1 Hyperion sensor was the first satellite to collect hyperspectral data from space (November 2000). Since then, the two major hyperspectral sensors developed are AVIRIS (Airborne Visible Infrared Imaging Spectrometer), which collects data in 224 bands, and the CASI-2 (Compact Airborne Spectrographic Imager-2), which has a spectral resolution of 228 bands. CHRIS (Compact High Resolution Imaging Spectrometer) mounted on the PROBA satellite is a new generation sensor that bridges the gap between imaging spectroscopy and multispectral sensors. CHRIS/PROBA combines the advantages of satellite platforms (homogeneous quality long-term data suitable for highly automated processing) with a relatively high spatial and spectral resolution, intermediate between ocean colour satellite sensors and hyperspectral airborne sensors. It has 63 spectral bands with a minimum bandwidth of 1.3 nm.

The major benefits of hyperspectral imaging are that data can be acquired anywhere, globally, at low cost to the end user, and that spaceborne sensors can provide year-round temporal data. Spaceborne sensors have a well defined sun-synchronous orbit, ensuring consistent illumination characteristics that can discern minute details of the surface features of the earth.

The spaceborne hyperspectral products can include geocoded maps of forest biomass and, more specifically, maps of above-ground carbon. To establish long-term trends as they relate to biomass and carbon budgets, change detection maps quantifying afforestation, deforestation and reforestation could be generated. The main benefits to end-users are that hyperspectral data can provide more detailed and accurate forest inventory information, and the ability to provide specialized products such as above-ground carbon maps. However, for change detection, hyperspectral data may not offer many improvements over multispectral data, since the types of change commonly being mapped are related to differences in canopy cover that fall within the capacity of a multispectral sensor's resolution.

Aerial videography

Videography, in the simplest sense, involves taking continuous overlapping frames of an area to record image data in analogue form on magnetic videotape. In the past decade, videography has become an increasingly popular tool for mapping. Broader application in recent years has resulted from an increase in the quality and availability of video equipment, and the advent of video analysis techniques and GPS, which permit multiple flightlines to be georeferenced. As a result of these developments, current applications are quite diverse in a variety of generalized land use inventory, monitoring and environmental studies. Graham (1993) reports on the use of video integrated with GPS to collect base data for vegetation mapping throughout Arizona. The major advantages of using aerial videography are as follows:

- acquisition of data is fast;
- coverage can be easily planned, directed and revised;
- all equipment for implementation is portable and does not require special aircraft;
- flightlines can be based on predetermined coordinate files;
- very high spatial resolutions are achievable;
- visual, infrared and thermal imaging can all be carried out;
- in-flight monitoring ensures complete cover and minimum flying time;
- data is replicable for monitoring purposes;
- surveys are a fraction of the cost of conventional aerial photography;
- low flying height and flexible imaging parameters minimize stand-by time;
- results may be easily imported and processed in image processing and GIS systems.

Brown et al (2005) have created a virtual tropical forest using a multispectral three-dimensional aerial digital imagery system (M3DADI) to develop a virtual forest where individual trees can be counted and crown areas and heights of all plant groups can be measured. The system collected high-resolution, overlapping stereo imagery through which carbon stocks in above-ground biomass for the pine in Belize could be estimated. A total of 77 plots were established on the images and, using a series of nested plots, the crown area and heights of pine and broadleaf trees, palmettos and shrubs were digitized. Based on standard destructive harvest techniques, highly significant allometric regression equations between biomass carbon per individual and crown area and height were obtained. The coefficient of variation was high for all vegetation types (with a range of 31–303 per cent, reflecting the highly heterogeneous nature of the system. The estimates of cost effectiveness of the technique showed that the conventional field approach took about three times more person-hours than the M3DADI approach.

Another major effort in utilizing videography was in the Gap Analysis Programme (GAP) of the United States Geological Survey (USGS) for mapping biodiversity across the United States, where large-scale georeferenced aerial videographic images were used as an alternative approach to ground truth verification.

Comparison of different techniques

Finding an optimum combination of accuracy of measurements and the cost of the technology is often a major challenge in projects using remote sensing for estimating forest or plantation biomass and carbon sequestration. To perform a comparative analysis of different remote sensing techniques for biomass estimation, the following factors need to be considered.

Measurement requirements
The measurement types involved in carbon sequestration and biomass studies can be changes in land use and land cover, in biomass and in the growing stock of biomass. Land use and land cover are usually defined as the aerial extent of discrete classifications whereas biomass density is a continuous variable averaged over the area.

Resolution requirements
The resolution requirements for biomass estimation and sequestration studies have been extensively studied. Townsend and Justice (1988) have estimated the resolution requirements for studies on land transformations and their study suggests a minimum resolution of 500 m. There has been tremendous technological progress since this study, and further studies have concluded that a resolution of 100 m or lower is the minimum requirement for biomass

estimation. There is a range of resolutions available that vary according to the sensor capabilities, antenna length, aperture size, transmitting power and orbital characteristics. Optical instruments have a better resolution than radar instruments. However, it is necessary to determine the resolution required for projects attempting to measure carbon sequestration or estimate biomass.

Coverage requirements

The coverage requirements of a study depend on the regions where the biomass needs to be measured and the proportion of the area to be measured. The requirements for repeating the measurements also need to be considered. Adequate sampling can possibly generate rough estimates at national levels for carbon stocks, but any incorrect extrapolations can cause huge uncertainties. Hence, for biomass estimation and carbon sequestration projects it is desirable to have 100 per cent coverage. There are also important issues of latitudinal and longitudinal coverage as well as the size of the footprint, which is inversely proportional to the ground resolution.

The availability of a signal also affects coverage, with passive systems needing reflected energy from the sun to obtain imagery, whereas infrared signals can be obtained at anytime. Dawn and dusk are considered the best times for obtaining signatures of interest. Cloud cover is another aspect that prevents coverage by causing reflection during the day. However, laser ranging and radar estimations may be carried out at anytime during the day without cloud cover.

Accuracy requirements

The degree of accuracy required will be dependent on whether the objective is the estimation of changes in area under forests or plantations, the estimation of biomass growing stock or growth rate, the estimation of total biomass production or the loss of biomass. Acceptable accuracy values could range from 10 to 30 per cent depending on the degree to which the biomass changes have to be measured rather than detected (Vincent, 1998).

For detecting change in land use, optical or infrared technologies are a better option. However, a combination of SAR and optical systems provides the maximum accuracy and the largest number of classes. Detection of biomass change is considered best by SAR data.

Remote sensing provides the opportunity for systematic and repetitive observations of large areas on a long-term basis. Furthermore, estimations in biomass change, standing biomass and carbon sequestration rates are also possible. In many cases, repetitive (annual and/or seasonal) measurements are generally required to detect changes in area and growth, particularly in rapidly changing environments. However, regional to global observations using fine spatial resolution data demand time and resources. A more efficient approach is to utilize coarser spatial resolution sensors with shorter revisit cycles and larger swaths for identifying 'hot spots' of change. Currently available sensors can acquire data on regional/global scales, but smoke, haze and other environmental factors could deter the applicability of such data. An increased number of measurements are also needed in order to address the hindrances from environmental parameters.

Feasibility of remote sensing techniques for Clean Development Mechanism (CDM) and bioenergy projects

Remote sensing is a continually improving science, beginning with a well designed data collection and data preparation activity and collecting the right data to answer well defined questions. There are a number of forest monitoring techniques and remote sensing compares well with traditional techniques to address the challenging questions of monitoring and evaluation of forestry and bioenergy projects. Remote sensing cannot be used effectively for below-ground biomass, deadwood, litter and soil organic carbon and can therefore focus only on above-ground biomass. A comparison of different techniques is given in Table 6.4.

Satellite remote sensing allows for synoptic, historic, actual and repeated views of the land use and vegetation in and around the project area which makes it a powerful tool to assess land use change processes and to monitor biomass or carbon sequestration.

Project development

Accurate measurements of biomass stock changes in project areas can be made using radar remote sensing. Historical time-series data can be used to assess changes in land use patterns. Once a biomass or carbon sequestration project is under way, all its land use activities can be documented through the use of high-resolution data.

Among the remote sensing techniques available, estimation of biomass and detection of biomass change can be best achieved using SAR data. Studies have shown that SAR data can detect the removal of half the tree trunks by selective logging. Modelling for biomass production or carbon sequestration, however, needs both optical and SAR data to be combined. Changes in land use can be obtained only by optical data. Remote sensing has developed into an important tool for monitoring and evaluation, and as a subsequent decision-making tool for various bioenergy and carbon sequestration projects. The feasibility of the use of remote sensing for monitoring of carbon stocks and flows in a project has been assessed by Vine et al (1999). The conclusive comparisons of various methods and their areas of applicability have been outlined in Table 6.5.

Table 6.4 *Advantages and disadvantages of forestry monitoring techniques*

Techniques	Advantages	Disadvantages
Modelling	• Relatively quick and inexpensive • Useful for baseline development • Can be used for bioenergy projects • Most useful as a complement to other methods	• Relies on highly simplified assumptions • Needs to be calibrated with on-site data
High-level remote sensing (satellite imagery)	• Provides relatively rapid regional-scale assessments of land cover, land use and green vegetation biomass • Useful for monitoring leakage	• Time and knowledge needed to transform spectral classifications into accurate land use or land cover classifications • Access to high quality imagery may not be available during certain seasons or due to sun angles • Has not been used to measure carbon. Can be quite expensive
Low-level remote sensing (aerial photography)	• Complements high-level remote sensing • Useful for monitoring leakage	• In test phase • Less expensive than high-level remote sensing
Field/site measurements	• Useful for determining what was actually implemented in projects and for tracking fate of wood products • Flexible in selection of methods and precision • Peer reviewed and field tested systems available • Using control plots, can calculate net carbon sequestration	• May be more expensive than other methods

Source: Vine et al (1999).

The complete project can be broken down into two phases: design and development, and implementation. During the design and development phase, primary information on forest cover and historical forest cover is needed, which can be covered with medium to moderate resolution data. This information can also be used at later stages, if combined with high-resolution data. During the implementation phase, a detailed monitoring of biomass or carbon in forests or plantations is required for the duration the project. Information is required for the determination of forest area, structure, biomass or carbon stock density and land use assessment. Remote sensing has been recognized as a useful tool

Table 6.5 *Forestry monitoring methods by forestry project type*

Methods	Carbon conservation		Biomass production or carbon sequestration and storage		Carbon substitution
	Small project	Large project	Small project	Large project	
Modelling	+	+	+	+	−
Remote sensing	−	+	−	+	+
Field/site measurements	+	+	+	+	+

Notes: + = applicable; − = not applicable.
Source: Vine et al (1999).

for optimizing the sampling scheme by stratification of the project area and in providing spatial information for cost-effective and accurate monitoring of biomass or carbon relevant indices. The end products would be analogue and digital maps compatible with a GIS.

FUTURE DIRECTIONS

Remote sensing techniques will increasingly be used in bioenergy and carbon sequestration projects for estimation, monitoring and verification of biomass production or carbon sequestration. There is a need to promote the application of remote sensing techniques, particularly for small-scale projects covering a few thousand hectares. There is also a corresponding need to reduce the uncertainty of estimates and cost of estimation, particularly for bioenergy or carbon sequestration projects.

As remote sensing is an ever-growing science, some potential areas for future research to expand the role of remote sensing across different bioenergy and carbon sequestration projects are as follows:

- Optical and SAR data fusion needs to be explored. While both optical and microwave data have their respective advantages and disadvantages, the fusion of such data holds great promise for biomass estimation.
- Studies on interferometric, polarimetric and/or multi-frequency SAR applications are necessary.
- Radar and lidar capabilities need to be enhanced and tested for three-dimensional imaging of forests.
- Field measurements and networking of global scale databases of ground truth and standardized biomass estimation models are necessary.

- There must be greater utilization of ground, aircraft and satellite remote sensing instruments to measure variables that describe the temporal and spatial dynamics of ecosystems (including CO_2 exchange, biomass, leaf area index), including the human impacts on these systems.
- Establishment of satellite systems with repetitive regional coverage for systematic data acquisition is required.
- Quantification of the uncertainties associated with biomass calculations and optimization of remote sensing techniques to reduce uncertainties required.
- Considerations of issues of accessibility and affordability of data should be addressed at global and particularly project scales.

ACKNOWLEDGEMENTS

The authors wish to thank the Ministry of Environment and Forests for their support of the Centre for Ecological Sciences in conducting research relevant to climate change and forests.

NOTE

1   The authors are affiliated with the Centre for Sustainable Technologies and Center for Ecological Sciences, Indian Institute of Science, Bangalore, India, e-mail: ravi@ces.iisc.ernet.in

REFERENCES AND FURTHER READING

Brown, S., Iverson L. R., Prasad, A. and Liu, D. (1993) 'Geographic distribution of carbon in biomass and soils of tropical Asian forests', *Geocarto International*, vol 8, no 4, 45–59

Brown S., Pearson T., Slaymaker D., Ambagis S., Moore N., Novelo D., Sabido W., (2005) 'Creating a Virtual Tropical Forest from Three Dimensional Aerial Imagery to Estimate Carbon Stocks', *Ecological Applications*, vol 15, no 3, 1083–1095

Fazaka, Z., Nilsson, M. and Olsson, H. (1999) 'Regional forest biomass and wood volume estimation using satellite data and ancillary data', *Agricultural and Forest Meteorology*, 98–99, 417–425

Foody, G. M., Boyd, D. S. and Cutlerc, M. E. J. (2003) 'Predictive relations of tropical forest biomass from Landsat TM data and their transferability between regions', *Remote Sensing of Environment*, vol 8, 463–474

Graham, L. A. (1993) 'Airborne Video for Near-Real Time Vegetation Mapping', *Journal of Forestry*, vol 8, 28–32; *Journal of Applied Ecology*, (2003), vol 40, 592–600

Jensen, J. R. (2003) *Remote Sensing of the Environment, An Earth Resource Perspective*, Pearson Educational Inc., Indian reprint

Kurvonen, L., Pulliainen, J. and Hallikainen, M. (2002) 'Active and passive microwave remote sensing of boreal forests', *Acta Astronautica*, vol 51, no 10, 707–713

Lefsky, M. A., Harding, D., Cohen, W. B., Parker, G., and Shugart, H. H (2005) 'Surface lidar remote sensing of basal area and biomass in deciduous forests of Eastern Maryland, USA', *Remote Sensing of Environment*, vol 67, 83–98

Le Toan, T., Beaudoin, A., Riom, J. and Guyon, D. (1992) 'Relating forest biomass to SAR data', *IEEE Transactions on Geoscience and Remote Sensing*, vol 30, no 2, 403–411

Mauro, G. (2003) 'High resolution satellite imagery for forestry studies: the beechwood of the Pordenone Mountains (Italy)' www.isprs.org/istanbul2004/comm4/papers/502.pdf

Nelson, R. F., Krabill, W. B. and Tonelli, J. (1988) 'Estimating forest biomass and volume using airborne laser data', *Remote Sensing of Environment*, vol 15, 201–212

Pulliainen, J., Engdahl, M. and Hallikainen, M. (2003) 'Feasibility of multi-temporal interferometric SAR data for stand-level estimation of boreal forest stem volume', *Remote Sensing of Environment*, vol 85, 397–409

Ranson, J. K. and Sun, G. (1994) 'Northern forest classification using temporal multi-frequency and multipolarimetric SAR images', *Remote Sensing of Environment*, vol 47, no 2, 142–153

Ranson, K. J. and Sun, G. (1997) 'An evaluation of AIRSAR and SIR-C/X-SAR data for estimating northern forest attributes', *Remote Sensing of Environment*, vol 59, 203–222

Ravindranath, N. H. and Hall, D. O. (1995) *Biomass, Energy and Environment: A Developing Country Perspective from India*, Oxford University Press

Read, J. M., Clark, D. B., Venticinque, E. M. and Moreira, M. P. (2003) 'Methodological insights application of merged 1 m and 4 m resolution satellite data to research and management in tropical forests', *Journal of Applied Ecology*, 40, 592–600

Roy, P. S. and Ravan, S. A. (1996) 'Biomass estimation using satellite remote sensing data: an investigation on possible approaches for natural forest', *Journal of Biosciences*, vol 21, no 4, 535–561

Sader, S. A. (1987) 'Forest biomass, canopy structure, and species composition relationships with multipolarization L-band synthetic aperture radar data', *Photographgrammetric Engineering and Remote Sensing*, vol 53, no 2, 193–202

Sandra, B., Pearson, T., Slaymaker, D., Ambagis, S., Moore, N., Novelo, D. and Sabido, W. (2004) 'Creating a virtual tropical forest from three-dimensional aerial imagery to estimate carbon stocks', *Ecological Applications*, vol 15, no 3, 1083–1095

Santos, J. R., Freitas, C. C., Araujo, L. A., Dutra, L. V., Mura, J. C., Gama, F. F., Soler, L.S. and Sant'anna, S. J. S. (2003) 'Airborne P-band SAR applied to the above-ground biomass studies in the Brazilian tropical rainforest', *Remote Sensing of Environment*, vol 87, 482–493

Tickle, P., Witte, C., Danaher, T. and Jones, K. (1998) 'The application of large-scale video and laser altimetry to forest inventory', *Proceedings of the 9th Australasian Remote Sensing and Photographgrammetry Conference*

Townsend, J. R. G. and Justice, C. O. (1988) 'Selecting the spatial resolution of satellite sensors required for global monitoring of land transformation', *Int. J. Remote Sensing*, vol 19, no 2, 187–236

Vincent, M. A. (1998) 'Scoping the potential of using remote sensing to validate the inclusion of carbon sequestration in international global warming negotiations',

Proposal submitted in response to the Terrestrial Ecology and Global Change (TECO) Research Announcement: NRA: 97-MTPE-15

Vine, E., Sathaye, J. and Makundi, W. (1999) *Guidelines for the Monitoring, Evaluation, Reporting, Verification, and Certification of Forestry Projects for Climate Change Mitigation*, Environmental Energy Technologies Division, March, Ernest Orlando Lawrence Berkeley National Laboratory

Wang, Y., Kaisischke, E. S., Melack J. M., Davis, F. W. and Christensen, N. L. (1995) 'The effects of changes in forest biomass on radar backscatter from tree canopies', *International Journal of Remote Sensing*, vol 16, 503–513

Wulder, M. A. and Franklin, E. (2003) *Remote Sensing of Forest Environments: Concepts and Case Studies*, Kluwer Academic Publishers

Case Studies

INTRODUCTION

Case studies are used to illustrate step-by-step methods for calculating biomass resources and uses in modern applications, based on fieldwork experiences, or to illustrate perceptions of potential major changes or trends in a particular area, for example the increasing use of bioenergy in modern applications. There are five case studies, individually authored, as detailed below.

- 7.1 International biotrade, which examines the development of international biotrade in bioenergy and its wider implications; this is, of course, a new trend which could have major impacts in supply and demand of bioenergy.
- 7.2 This case study looks very briefly at how to build a modern bioenergy market by examining how it was created in Austria over the years. It shows how a combination of favourable factors, for example, availability of resources (raw material, human capital, local participation, financial), are necessary for establishing a successful modern and efficient bioenergy market.
- 7.3 Deals with biogas use in small island communities and is illustrative of small-scale and traditional ways of dealing with bioenergy applications.
- 7.4 This case study examine the utilization of biodiesel from coconut and jatropha in small-scale applications; this is an attractive option, particularly in remote locations and small markets.
- 7.5 The last case study looks at the potential role of biomass in carbon sequestration and climate change, a fundamental area of concern to us all. It highlights the benefits of bioenergy, which can be very large or unnoticeable in respect of potential impacts on climate change, depending on the total amount used and how we utilize bioenergy on a global scale.

CASE STUDY 7.1   INTERNATIONAL BIOTRADE:
POTENTIAL IMPLICATIONS FOR THE DEVELOPMENT
OF BIOMASS FOR ENERGY

Frank Rosillo-Calle

Unlike fossil fuels, biomass for energy has hardly been traded regionally or nationally, and even less so internationally. Most trading of bioenergy has occurred at local level. But this is now changing and biomass for energy is rapidly becoming a major traded commodity, particularly as densified biomass (i.e. pellets and chips) and liquid biofuels (i.e. ethanol and biodiesel). This new phenomenon could have major implications for the development of bioenergy, both the potential benefits and possible pitfalls. Case study 7.2 further illustrates how a modern biomass energy market was created by examining the case of Austria.

The rapidly growing trend in international biomass trade was recognized by the IEA which subsequently created 'Bioenergy Task 40: Sustainable International Bio Energy Trade: Securing Supply and Demand', on which most of the information contained in this case study is based (see www.bioenergytrade.org). Task 40 was established in December 2003, and has as its prime focus to assess the implications of international biotrade. Task 40 is likely to be a pivotal task in the international development of biomass for energy.

The rationale for biotrade

Over the past decades, the modern use of biomass has increased rapidly in many parts of the world. Given the important role that biotrade is likely to play in the rapidly expanding international bioenergy sector, both positive and negative outcomes are highly possible. It is important to be aware of the potential outcomes.

In a world where the demand for energy seems endless, bioenergy is bound to pay a critical role. A reliable supply and demand is vital to develop stable market activities, aimed at the bioenergy trade. Given the expectation for high global demand on bioenergy, the pressure on available biomass resources will increase. Without the development of biomass resources (for example, through energy crops and better use of agro-forestry residues) and a well functioning biomass market to assure a reliable and lasting supply, those ambitions may not be met or, worse, they could have serious negative impacts on the sustainability of supply.

The development of truly international markets for bioenergy may become an essential driver to develop the potential of bioenergy, which is currently underutilized in many regions of the world. This is true for both residues, for dedicated biomass energy plantations (forestry and crops) and for multifunctional systems such as agro-forestry. The possibility of exporting biomass-derived commodities for the world's energy market can provide a stable and reliable demand for rural communities, thus creating important socioeconomic development incentives and market access. A key objective is to ensure that biomass is produced in a sustainable way.

Opportunities and pitfalls of biotrade

The development of an international bioenergy trade will be unequal around the world because there are many regions endowed with poor resources. For some developing countries endowed with large biomass resources and lower costs, biotrade offers a real opportunity to export bioenergy primarily to developed countries. Many developing countries have a large potential for technical agro-forestry residue as well as for dedicated energy plantations (e.g. ethanol from sugar cane, pellets or charcoal from eucalyptus plantations), and this could represent a good food business opportunity.

The possibilities to export biomass-derived commodities to the world's energy markets can provide a stable and reliable demand for rural communities in many (developing) countries, thus creating an important incentive and market access in many rural areas.

International biotrade is still very small, but growing rapidly; for example, in Europe it amounts to approximately 50 PJ per year (mainly wood pellets and forest residues). Many trade flows are between neighbouring countries, but long-distance trade is also increasing. Examples are the export of ethanol from Brazil to Japan and the EU, palm kernel shells (a residue of the palm oil production process) from Malaysia to the Netherlands, wood pellets from Canada to Sweden.

In the short to medium term, the most promising areas for international trading are

- woodchips and other densified biomass (briquettes and pellets) for CHP and co-firing in coal power plants;
- ethanol fuel;
- charcoal (for example, in 2000 Brazil exported about 8000 t of charcoal, worth US$1.4 million);
  briquettes, traded in value-added markets (e.g. restaurants).

Ethanol seems to be one of the most realistic options for setting up a truly global bioenergy trade, at least within the present decade. Use of ethanol fuel is growing rapidly as it has a considerable potential for substituting oil in the transportation sector, given the right conditions.

Currently, over 3 billion litres of fuel ethanol is traded annually, with Brazil and the United States being the main exporters, and Japan and the EU the main importers. However, this is expected to increase dramatically as international demand is growing rapidly, with more than 30 countries interested in ethanol fuel. One of the major constraints is that few countries, excluding Brazil and a few others, are in a position to supply both their growing domestic market and international demand. For example, Japan, as well as other potential importers, would like to import far more but are constrained by the high inelasticity of the ethanol fuel market. If the ethanol fuel market becomes large enough, it could create sufficient liquidity to attract many ethanol fuel players. A large ethanol market would be able to guarantee to cover the possibility of any shortfall in any one country or region, as is the case with oil and gas.

These trade flows may offer multiple benefits for both exporting and import-ing countries. For example, exporting countries may gain an interesting source of additional income and an increase in employment. Also, sustainable biomass production will contribute to the sustainable management of natural resources. Importing countries on the other hand may be able to fulfil their GHG emission reduction targets cost effectively and diversify their fuel mix.

For interested stakeholders (e.g. utilities, producers and suppliers of biomass for energy), it is important to have a clear understanding of the pros and cons of biomass energy. Investment in infrastructure and conversion facilities requires risk minimization of supply disruptions, in terms of volume and quality as well as price.

The long-term future of large-scale international biotrade must rely on environmentally sustainable production of biomass for energy. This requires the development of criteria, project guidelines and a certification system, supported by international bodies. This is particularly relevant for markets that are highly dependent on consumer opinions, as is presently the case in western Europe.

What are the main drivers for international biotrade?

Clear criteria and identification of promising possibilities and areas are crucial as investments in infrastructure and conversion capacity rely on minimization of risks of supply disruptions. A summary of the main emerging drivers in inter-national biotrade (see www.bioenergytrade.org) follows:

1   *Cost effective GHG emission reduction.* At present, climate policies in various countries, such as the EU, are major factors in the growing demand for bioenergy. In situations where indigenous resources are insufficient or costs are high, imports can be more attractive than exploiting local biomass poten-tials. This will change over time as costs are reduced, but several world regions will continue to have inherent advantages in producing lower cost biofuels, particularly tropical developing countries.

2   *Socio-economic development.* There is ample evidence indicating a strong positive link between developing bioenergy use and local (rural) develop-ment. Furthermore, for various countries bioenergy exports may provide substantial benefits for trade balances.

3 *Fuel supply security.* Biomass for energy will diversify the energy mix and thus prolong the lifespan of other fuels (fossil fuels), therefore reducing the risks of energy supply disruptions. This argument is particularly strong in the case of liquid biofuels (ethanol and biodiesel) as the transport sector is overwhelmingly dependent on oil.

4 *Sustainable management and use of natural resources.* Large-scale production and use of biomass for energy will inevitably involve additional demand on land. However, if biomass production can be combined with better agricultural methods, restoration of degraded and marginal lands and improved management practices to ensure that it is produced sustainably, this in turn can provide a sustainable source of income for many rural communities.

Are there any particular barriers to international biotrade?

Based on a literature review and interviews, a number of categories of potential barriers have been identified. These barriers may vary a great deal in terms of scope, relevance for exporting and importing countries and now stakeholder perception. A summary of the main barriers is given in the following paragraphs.

Economic barriers

One of the principal barriers for the use of biomass energy in general is the competition with fossil fuel on a direct production cost basis (excluding externalities). There are many hidden costs of fossil fuels which are not reflected in the market.

Many governments around the world have now introduced various mechanisms (legislation, subsidies, compulsory purchasing, etc.) to promote bioenergy, particularly for electricity, heat and mandatory blending in transportation (ethanol and biodiesel). However, often such support is insufficient, particularly when it comes to long-term policies, which in turn discourages long-term investment as bioenergy is still often considered too risky by many investors. This problem is further compounded by a lack of harmonization at many levels (e.g. policy among EU member states, standards, etc.).

Technical barriers caused by physical–chemical characteristics of bioenergy

This is a major difficulty that those dealing with bioenergy have to come to terms with. While markets are gradually accepting and adapting to bioenergy, some major barriers still remain. For example, certain physical and chemical properties such as low density, high ash and moisture content, nitrogen, sulphur or chlorine content, etc., make it more expensive to transport, and often unsuitable for direct use (e.g. in co-firing power plants). Overcoming these difficulties will require many technical improvements (e.g. to boilers) and changes in attitudes of major users.

Logistical barriers

An important limiting factor to trading biomass is that it is often bulky and expensive to transport. One solution is to compact it to make it economical and practical to move long distances at acceptable costs. Fortunately densification technology has improved significantly recently and, as detailed in Appendix 4.2 of Chapter 4, densified biomass is already being commercialized on a large scale. High-energy liquid biofuels, such as ethanol, vegetable oils and biodiesel, do not present such difficulties and can easily be transported cheaply over long distances.

Large-scale transportation of bulky biomass is just beginning to take off and thus our experience is very limited. For example, very few ships exist today that are specially adapted for this purpose and this will probably result in higher transportation costs. However, some studies have shown that long-distance international transport by ship is feasible in terms of energy use and transportation costs; but availability of suitable vessels and the possibility of adverse meteorological conditions (e.g. during the wintertime in Scandinavia and Russia) need be considered.

Harbours and terminals that do not have the capacity to handle large biomass streams can also hinder the import and export of biomass to certain regions. The most favourable situation is when the end-user has the facility close to the harbour, avoiding additional transport by trucks. The lack of significant volumes of biomass can also hamper logistics. In order to achieve low costs, large volumes need to be shipped on a more regular basis.

International trade barriers

International biotrade is a fairly recent development and thus often no specific biomass import regulations exist, which can be a major hindrance to trading. For example, in the EU most residues that contain traces of starches are considered potential animal fodder, and thus subject to EU import levies.

Although not exclusive to bioenergy, the potential contamination of imported biomass material with pathogens or pests (e.g. insects or fungi) can be a major impediment. For example, roundwood imported into the EU can currently be rejected if it seems to be contaminated; as will be the case with any other type of biomass.

Ecological barriers

Large-scale dedicated energy plantations also pose various ecological and environmental problems that cannot be ignored, ranging from monoculture, long-term sustainability, potential loss of biodiversity, soil erosion, water use, nutrient leaching and pollution from chemicals. However, to retain a sense of perspective, biomass for energy is no better or worse than any other traded community.

Social barriers
Large-scale energy plantations also have potentially major social implications, both positive and negative; for example, the effect on the quality of employment (which may increase or decrease, depending on the level on mechanization, local conditions, etc.), potential use of child labour, education and access to health-care, etc. However, such implications will reflect prevailing situations and would be no better or worse than in relation to any other similar activity.

Land availability, deforestation and potential conflict with food production
This issue has been discussed already (see Chapter 1). Food versus fuel is a very old issue that refuses to ago away, despite the fact that a large number of studies have demonstrated that land availability is not the real problem. Food security should not be affected by large energy plantations if proper management and policies are put in place, and government policy and market forces would favour food production rather than fuel in case of any food crisis. However, food availability is generally not the problem, rather the lack of purchasing power of the poorer members of the population to purchase it. In developed countries, where there is surplus land, the main issues may be competition with fodder production rather than food crops and higher cost of food.

Methodological barriers – lack of clear international accounting rules
This can be a serious problem, at least in the short term, since clear rules and standards need to be established before international biotrade can be developed. The nature of biomass can also pose problems for trade because it can be considered as a direct trade of fuels and also as indirect flows of raw materials that end up as fuels in energy production during or after the production process of the main product. For example, in Finland the biggest international biomass trade volume is indirect trade of roundwood and wood chips. Roundwood is used as a raw material in timber or pulp production. Wood chips are raw material for pulp production. One of the waste products of the pulp and paper industry is black liquor, which is used for energy production.

Legal barriers
Biotrade is further complicated by the fact that each country, even countries within the EU, has its own legislation to deal with international trade. For example, emission standards differ significantly from country to country and this is an area that affects biotrade because the potential exists for carbon credits, for example. Therefore, common emission standards among major trading blocs will result in important positive benefits for international biotrade.

So what lessons can be learned from biotrade?

This case study has raised, rather than answered, a number of important issues. For example, some questions that need to be asked are, should biotrade be treated differently from any other commodity, given its nature? Should market forces be left to determine supply and demand as with any other commodity? Is biotrade so special that it needs to be considered in its own right? How much control should be exercised over biotrade and will this enhance or hinder its development? Can international biotrade distort domestic supply? What does this potentially mean for traditional biomass applications? What would be potential impacts on food, particularly prices?

Further reading

www.bioenergytrade.org (or old site: www.fairbiotrade.org). This site provides you with a lot of information on biotrade (objectives of Task 40, workshops, documents, country reports, publications, contacts, etc.).

Case study 7.2   Building a modern biomass energy market: The case of Austria

Frank Rosillo-Calle

Case study 7.1 made the case for international bioenergy trade. This case study looks at Austria and the reasons why bioenergy has been so successful there and why it plays such an important role in the country's energy system. There are many important lessons to be learned from the Austrian experience. This is particularly so because it is used at all levels: from small to large and from domestic to industrial uses.

The Austrian case is more interesting if one takes into account the fact that the success of bioenergy took place against a backdrop of low energy prices (especially oil) and at a time when very little R&D was being channelled into non-traditional energy alternatives. In the current energy climate it would be much easier to establish a bioenergy market, not only because of the cumulative effects on the market worldwide (e.g. improved conversion technologies, know-how, political support, etc.), but also because the costs would be lower by comparison.

In 2002, approximately 12 per cent of Austria's primary energy consumption originated from biomass. The most significant contribution of bioenergy is in the domestic sector (individual households) which represents 60 per cent; followed by process heat generation with 21 per cent; CHP and thermal power plants with 11 per cent; and district heating with 8 per cent (see www.energyagency.at). An additional factor which makes Austria particularly illustrative is the high efficiency of bioenergy applications; for example, biomass plant efficiency has increased from an average of 50 per cent in 1980 to over 90 per cent today while at the same time CO_2 emissions have decreased to less than 100 mg/Nm^3.

So why has Austria been so receptive to bioenergy? Obviously this is due to a combination of factors that may, or may not, be repeated elsewhere else. The following seem to be key:

- availability of natural resources;
- long-term political commitment;
- active participation of industry and individuals;
- the generally positive attitude of the Austrians, which in turn could be partly explained by the fact that Austrians prefer to live in individual dwellings (single-family houses) rather than flats;
- innovative attitude (the bioenergy industry is full of individual innovators); this might tell us a lot about educational standards;
- availability of cheap capital/financial resources;
- availability of long-term R&D funding.

Market penetration

Markets do not operation in a vacuum; they need to be built up over time. In Austria, six large bioenergy markets have been successfully established:

- forest related industries, since the 1950s;
- district heating of rural villages and towns, since 1980;
- medium-scale biomass heating projects (schools, town halls, village communities, etc.), since 1992;
- wood pellet heating for single-family houses, since 1995;
- co-generation of heat and power, since 2002;
- biogas production from energy crops, since 2002.

The main drivers that led to increasing usage of bioenergy in Austria can be summarized:

- availability of biomass (e.g. 47 per cent of the territory is covered by forest);
- a long tradition of bioenergy (wide portfolio of proven technologies, good information and positive image);
- long-term political commitment;
- attractive framework conditions (e.g. stable and predictable financial incentives, fuel price stability, commercially driven, high quality of appliances);
- political cooperation at national, regional and local levels when it comes to solving problems.

But often even these favourable circumstances might not have been sufficient to create a flourishing bioenergy market. In Austria the drivers to success for electricity generation from wood, for example, include:

- the Eco-Power Act of 2002, which created favourable legal framework conditions;

- the respective ordinance, introduced in January 2003, which provides for guaranteed tariffs for up to 13 years for plants licensed in 2004 entering into operation by mid-2006;
- increased investment activities which can have a snowball effect, encouraging yet more activities.

The availability of subsidized capital is also a key factor. This is a particularly acute problem in many developing countries, but Austria is a rich country and those involved in the bioenergy business have at their disposal large amounts of cheap, highly subsidized investment capital, for example:

- in the forest related industries, up to 30 per cent for the heat part of biomass CHP plants;
- 30–40 per cent of project costs for biomass district heating in rural villages and small towns;
- 30 per cent of project costs for medium-scale biomass heating projects (e.g. schools, community centres, etc.);
- up to 30 per cent of project costs for biomass boilers and ovens for single-family houses;
- up to 30 per cent for the heat parts of CHP co-generation.

Can this case be repeated elsewhere? Are these conditions necessary for other successful bioenergy programmes? Can poorer countries afford such programmes? To answer these questions will require a wider assessment of the whole energy system that takes into account the full costs of externalities. However, judging also by the rapid investment growth in renewable energy, including bioenergy, the market seems to be moving to maturity rapidly.

In the present circumstances there must be instruments to stimulate the market be they financial, political, etc. But judging by the Austrian case, other factors are also important to create a market such as a high standard of education, training of human resources and a good dissemination strategy.

References and further reading

For US-specific conversion factors visit http:/bioenergy.ornl.gov/papers/misc/energy_conv.html

www.energyagency.at

www.eva.wsr.ac.at

www.ieabioenergy.com/media/20_BioenergyinAustria.htm

Nemestothy, K. (2005) 'Biomass for Heating and Electricity Production in Austria' (www.energyagency.at)

Worgetter, M., Rathbauer, J. Lasselsberger, L., Dissemond, H., Kopetz, H., Plank, J. and Rakos, C. H. 'Bioenergy in Austria: potential, strategies, and success stories' (www.blt.bmlf.gv.at)

CASE STUDY 7.3   BIOGAS AS A RENEWABLE TECHNOLOGY OPTION FOR SMALL ISLANDS

Sarah Hemstock

Introduction

As indicated in Chapter 4 under 'Secondary fuels (liquid and gaseous)', biogas is an important source of energy, particularly in Asian countries such as China, India, Nepal and Vietnam. Although biogas is increasingly used in large industrial application, its primary use is in small-scale uses. It is of particular interest in small rural communities where there are few other alternatives. Biogas is an attractive alternative because it can provide various simultaneous benefits:

1   energy
2   sanitary
3   fertilizer.

This case study looks at the use of biogas in various small island communities in the South Pacific Islands.

Tuvalu is an independent constitutional monarchy in the south-west Pacific Ocean, located approximately 1000 km north of Fiji and with a total land mass of only 26 km², spread over 750,000 km² across its exclusive economic zone. It is composed of nine low-lying coral atolls, islands and numerous islets with the largest island covering only 520 ha and the smallest 42 ha. The average elevation is 3 m. The capital island, Funafuti, is only 2.8 km² but accounts for around half of the total population of approximately 11,000 and two-thirds of the gross domestic product (GDP) of the country.

The nation is regarded as exceptionally vulnerable to rising sea levels and increased storm activity as the maximum height above sea level is a mere 5 m. The climate is sub-tropical, with temperatures ranging from 28 to 36°C, uniformly throughout the year. There is no clear marked dry or wet season. The mean rainfall ranges between 2700 and 3500 mm per year but there are significant variations from island to island. Keeping pigs is a traditional activity and every household owns at least one pig (Rosillo-Calle et al 2003; Woods et al, 2005; Alofa Tuvalu, 2005) (see Table 7.1).

Tuvalu is close to being a totally oil-dependent economy. In 2004, total energy consumption was 4.6 ktoe. Of this, oil accounted for 3.8 ktoe (82 per cent) of total primary energy consumption and biomass accounted for 0.8 ktoe (almost 18 per cent of total primary energy consumption) (also see Chapter 5, Figures 5.2 and 5.3). This total includes the diesel charged by the two vessels (Nivaga II and Manu Folau) in Suva, Fiji (Hemstock and Raddane, 2005).

Table 7.1 *Privately occupied households by island/region and number of livestock*

Island/region	Number of households	Pigs	Chicken	Ducks	Cats	Dogs	Other
Funafuti	639	2,275	428	65	666	931	40
Outer islands	929	6,519	12,244	2,827	1,301	1,019	113
Tuvalu	1,568	8,794	12,672	2,892	1,967	1,950	153

Sources: Alofa Tuvalu Survey, 2005; Hemstock, 2005; Tuvalu 2002 Population and Housing Census, Volume 2: Analytical Report; Social and Economic Wellbeing Survey 2003, Nimmo-Bell & Co.

Energy consumption

Annual energy consumption is over 0.4 toe per capita (approximately one-tenth of someone living in the UK or France) (Hemstock and Raddane, 2005). Currently, all of Tuvalu's oil is imported – this creates a vulnerable position since increasing oil prices and a drop in global oil production mean that Tuvalu's economy is in the hands of the oil suppliers.

Kerosene usage

Domestic kerosene use in Tuvalu was estimated at 263 toe annually, which is considerable when compared with 170 toe for air transportation (Hemstock and Raddane, 2005).

The use of biogas as a fuel for cooking should reduce kerosene use. Biogas is produced by the anaerobic fermentation of organic material. Production systems are relatively simple and can operate at small and large scales practically anywhere, with the gas produced being as versatile as natural gas. This is a very significant technology option for Tuvalu since anaerobic digestion can make a significant contribution to disposal of domestic and agricultural wastes and thereby alleviate the severe public health and water pollution problems that they can cause. The remaining sludge can then be used as a fertilizer (providing there is no polluting contamination) and actually performs better than the original manure since nitrogen is retained in a more useful form, weed seeds are destroyed and odours reduced. In addition, it would reduce the amount of household income spent on cooking fuel, provide much needed compost for family gardens (seawater encroachment has contaminated traditional taro pits) and thereby provide additional food security, dispose of pig waste which is currently swilled into the lagoons, and may provide additional household income from pig rearing and sale of gas.

Additionally, the use of biogas systems in a country such as Tuvalu could increase agricultural productivity. All the agricultural residue, animal dung and human sewage generated within the community is available for anaerobic digestion. The slurry that is returned after methanogenesis is superior in terms of its nutrient content; the process of methane production serves to narrow the carbon:nitrogen ratio (C:N), while a fraction of the organic nitrogen is mineralized to ammonium (NH_4^+) and nitrate (NO_3^-), the form which is immediately available to plants. The resulting slurry has double the short-term fertilizer effect of dung, but the long term fertilizer effects are cut by half (Chanakya et al, 2005). However, in a tropical climate, such as Tuvalu's, the short-term effects are the most critical, as even the slow degrading manure fraction is quickly degraded due to rapid biological activity. Thus, an increase in land fertility could result in an increase in agricultural production. Value added benefits include improved subsistence, increased local food security and income generation from increased land productivity.

Small-scale digesters

Small-scale digesters are appropriate for small and medium-sized rural farms. Typical fixed-dome small-scale digester sizes range from 4–5 m³ total capacity design for small single-family farms to 75–100 m³ total capacity designs. A digester having a capacity of 100 m³ can handle about 1800 kg of manure per day and would be suitable for a farm with about 30 cows or 150 pigs. This scale would be most suited to Tuvalu Model Piggery – a community-owned piggery on Funafuti where people rent stalls to house their pigs.

Basic digester systems will produce around 0.5 m³ of biogas per m³ of digester volume. For a family of six, digester systems of size 4–6 m³ can meet daily biogas requirements (about 2.9 m³) for all residential and agricultural uses. Efficient digesters with gas recovery systems may reduce methane emissions up to 70 per cent, with larger reductions achievable for longer retention times. This technology can make small and medium-sized farms more self-sufficient and reduces GHG emissions.

In theory biogas could be produced from the unused pig waste (see Table 7.2). Using a conservative collection efficiency of 60 per cent of total dung produced and conservative conversion efficiencies from family-based 6 m³ digesters

Table 7.2 *Energy available for biogas digestion from pig waste in Tuvalu*

Total number of pigs in Tuvalu (head)	Total annual production of pig manure (t/yr)	Total annual production of energy from pig manure (GJ/yr)	Total amount of manure available annually for use in biogas digester[a] (t/yr)	Total amount of energy available annually for use in biogas digester[a] (GJ/yr)
12,328 (60% = 7,397)	3,600	32,400 (771 toe)	2,106	19,440 (463 toe)

Source: [a] Assumes 60 per cent collection efficiency as some waste is used for compost and some will be difficult to collect (Alofa Tuvalu, 2005; Hemstock, 2005).

(15 pigs per digester), a total of 1,578 m³ of gas could be produced per day – providing 13,236 GJ/yr (315 toe). This is more than enough to provide the cooking gas (and possibly electricity for lighting) for 526 households across Tuvalu. Additional benefits include compost production, cleaner pig pens and the removal of a smelly hazardous waste.

Obviously, establishing a community need for energy services is the first step in the planning process, which was achieved via a series of community meetings. In addition, the resources available must be the initial starting point for any community based bioenergy project. Failed projects show that communities must be involved in the decision making and project planning process from the outset (Woods et al, 2005). In order to ensure the sustainability of any intervention, women were to be involved from the outset as they would be the main users of biogas as a domestic fuel. A woman's meeting was held (in conjunction with 'Women's Week') in the main meeting hall on Vaitupu (17 August 2005) to discuss biogas technology and implementation. Over 100 women were in attendance. The women were very positive about the technology and decided unanimously that they were committed to the idea and wanted to explore the possibilities of implementing biogas on Vaitupu.

After training sessions and discussions it was decided that a planning committee of 21 women would take the idea forward for Vaitupu. Over the next few days they developed an implementation strategy involving a series of 15 family-sized (6 m³) digesters. They obtained permission to use land to site the family digesters and pig pens and also planned a larger community-sized digester for each of the two villages on Vaitupu, although no agreement was reached as to where to site the community plants. The women also decided that the promotion of family gardens would run alongside the implementation of the family-sized digesters.

Again, theoretically, the unused pig waste could be utilized for biogas production (see Table 7.3). As in the Tuvalu example given above, using a conservative collection efficiency of 60 per cent of total dung produced and conservative conversion efficiencies from family-based 6 m³ digesters (15 pigs per digester), a total of 263 m³ of gas could be produced per day, providing 2433 GJ/yr (58 toe).

Table 7.3 *Energy available for biogas digestion from pig waste in Vaitupu*

Total number of pigs on Vaitupu (head)	Total annual production of pig manure (t/yr)	Total annual production of energy from pig manure (GJ/yr)	Total amount of manure available annually for use in biogas digester[a] (t/yr)	Total amount of energy available annually for use in biogas digester[a] (GJ/yr)
2,267	662	5,957 (142 toe)	397	3,574 (85 toe)

Source: [a] Assumes 60 per cent collection efficiency as some waste is used for compost and some will be difficult to collect (Alofa Tuvalu, 2005; Hemstock, 2005).

This is equivalent to total kerosene use for cooking in all households in Vaitupu. It is more than enough to provide the cooking gas (and possibly electricity for lighting) for 91 households. Additional benefits include compost production, cleaner pig pens and the removal of a smelly hazardous waste from around dwellings (see also Appendix 2.1).

Building on the success in Vaitupu, a further series of eight women's group meetings were held in Funafuti with each of the outer islands' Women's Associations based in Funafuti and with a women's gardening association. Meetings with women's groups were held since Alofa Tuvalu works directly with communities and includes project beneficiaries from the planning process through to implementation. This approach should give beneficiaries a sense of ownership of the project and will help to ensure that any intervention and exit strategy is socially, environmentally and economically sustainable. Women's groups were approached initially since women are the main users of biogas and any activity concerning the implementation of this technology should involve them directly.

The basic format for each meeting was:

- to describe the background of Alofa Tuvalu and the 'Small is Beautiful' project;
- to make clear links between climate change, carbon emissions and energy use;
- to point out the benefits of family gardens (food security, improved income, reduce waste by compost making, reduced reliance on imported foods, etc.);
- to give instructions on how to set up a family garden (ground preparation, where to buy compost, planting seeds, watering, pollination, collecting and storing seeds, organic fertilizer and composting techniques, etc.);
- to supply seeds to the women (tomato, lettuce, basil, melon, marrow, chilli pepper);
- to describe biogas technology and types of implementation – using examples of community biogas plants in India and a pig farm in Suva, Fiji. Discussing the added benefits of a reliable gas supply and compost production. Providing women with information about the technology so that they can formulate an initial project strategy.

In general, the attendees were extremely positive about biogas and family gardens. The attendees requested help with compost making and marketing surplus home garden produce. However, there was some concern that Alofa Tuvalu was not operating through the correct channels. The women wanted to use the technology and thought the best way to implement a project would be through the Kaupule (local government). However, it was stressed that Alofa Tuvalu wanted to work directly with women's groups as women would be the main users of the technology. The women were enthusiastic about the idea and wanted to explore the possibilities of implementing biogas on Funafuti.

Nui Island Women's Association was the most enthusiastic of the groups approached in Funafuti. All the women were very interested in starting their own family gardens. They also picked up on the idea of installing biogas very quickly and thought that it would be beneficial to Nui as people keep many pigs and have land to site the digesters. Most of the Nui population do not own land in Funafuti.

For Funafuti, since the Nui women do not have any land, they came up with the idea of a national system – along the lines of the Suva example (see the 'Project feasibility analysis' below) with more pig housing than the Tuvalu Model Piggery (TMP). However, none of the attendees kept their pigs at TMP because it was built on land belonging to Funafuti people, so they felt 'discouraged from using it'. The women advised the project organizers that issues concerning land ownership would have to be dealt with at the outset.

The women wanted cooking gas, rather than electricity production, since the supply of bottled gas is erratic and expensive as it has to be imported from Fiji.

Using a conservative collection efficiency of 60 per cent of total dung produced and conservative conversion efficiencies from family-based 6 m³ digesters (15 pigs per digester), a total of 491 m³ of gas could be produced per day – providing 4116 GJ/yr (98 toe). This is more than enough to provide the cooking gas (and possibly electricity for lighting) for 153 Funafuti households (see Table 7.4).

For Nui, the attendees wanted to run a large piggery (150–200 pigs) and digester as a Women's Community Project. Land had been donated to the Women's Community on Nui and attendees thought that a project of this type would be a good use of the land. After much discussion the women decided on the following.

- The women should be trained in all aspects of digester running and maintenance as well as husbandry.
- That the women would take it in turns to tend their pigs and collect coconuts for the pigs and labour would be free.
- The Women's Community would own a proportion of the pigs, which would be reared and sold to the Tuvalu Cooperative Society shop to generate revenue.

Table 7.4 *Energy available for biogas digestion from pig waste in Funafuti*

Total number of pigs on Funafuti (head)	Total annual production of pig manure (t/yr)	Total annual production of energy from pig manure (GJ/yr)	Total amount of manure available annually for use in biogas digester[a] (t/yr)	Total amount of energy available annually for use in biogas digester[a] (GJ/yr)
3,834	1,119	10,075 (240 toe)	672	6,045 (144 toe)

Source: [a] Assumes 60 per cent collection efficiency as some waste is used for compost and some will be difficult to collect (Alofa Tuvalu, 2005; Hemstock, 2005).

- The use of the pig pens should be free to the women looking after the set-up and a hire charge made for those who were not participating in the work.
- Cheap gas would be available for members of the Women's Community and a higher charge for non-members;
- Cheap compost would be provided to members of the Women's Community and a would be available at a higher charge to non-members.
- A vegetable garden would be set up in combination with the digester where vegetables would be cheaper for members of the Women's Community and available at a higher charge to non-members.

The women were 'very excited' about the prospects for their planned project on Nui and were asking questions such as 'What can we feed the pigs to get them to produce more gas?'

Project feasibility analysis based on Colo-i-Suva Pig Farm, Fiji

Technical and resource feasibility
The project was deemed technically viable and assessment revealed that there were more than sufficient dung resources to supply the proposed digesters in Tuvalu.

Economic feasibility
Costs associated with a 6 m³ digester are detailed in Table 7.5 (the Colo-i-Suva farm digester was 20 m³).

Table 7.5 *Associated costs for a 6 m³ digester*

	Appliance	
Subtotal – pipes and fittings (£)	Low cost scenario 149.83	High cost scenario 237.73
	Construction materials	
	Cost range (£)	
Brick or stone	68.14	108.11
Sand	16.03	25.44
Gravel	6.41	10.17
Labour	24.05	38.16
Rod – 8 mm	4.88	7.74
Cement	76.95	122.10
Sub total	196.46	311.71
Total build cost	£346.29 to	£549.44

Biogas electricity generation equipment would be expected to cost a further £600 to £800, and drainage for pig pens £600. For Tuvalu, costs would be further increased by costs for sea transportation of materials.

- Assumptions:
    — the life of a biogas plant is assumed to be 25 years;
    — the cost of fuelwood in Fiji is assumed to be £0.03 kg^{-1};
    — maintenance costs of the digester are between £5.50 and £8.50 per year.
- Annual savings = £137.34:
    — fuel wood (7 kg per day @ 0.03 per kg) = £76.65;
    — kerosene/LPG = £18.00;
    — fertilizer (slurry) = £42.69.
- Annual costs = £42.46:
    — maintenance = £8.50;
    — labour = £18.96;
    — other costs = £15.00.
- Payback time = 3.6 to 5.8 years (this figure assumes no subsidies).

Anticipated implementation problems
Occasionally, after periods of severe rainfall, the Fiji digester filled with too much water and had to be drained. The farmer did not service the pipes running from the pig pens to the digester so they occasionally became blocked.

Problems such as rainwater flooding of the digester can easily be avoided by correct site preparation when the digester is built; other technical problems with the project arose because of improper maintenance. Training and an agreed maintenance schedule could overcome these problems.

There is currently no example of biogas technology being implemented on a coral atoll. Construction costs may be higher than anticipated due to the difficulties of a high water-table and seawater flooding.

Land tenure issues are complicated in Tuvalu so any land use issues must be worked through before construction commences.

For this type of project there is an important role for innovative financing to enable consumers to spread the high initial cost of energy conversion technology over the life of the equipment (see the cost analysis above). This is particularly true for poor farmers running smaller scale digesters who, by definition, have little to offer as collateral and who are unfamiliar with formal credit systems.

What then are the main lessons from this case study? It shows that biogas is a potential alternative in certain small communities. Another important lesson is that it is vitally important to adopt a bottom-up approach; that is, that those people (in this case women), who stand to benefit most are involved in the project from its inception.

References

Alofa Tuvalu (2005) Tuvalu field survey results: July–October 2005, Alofa Tuvalu, 30 rue Philippe Hecht, 75019 Paris, France

Chanakya, H. N., Svati Bhogle and Arun, A. S. (2005) 'Field experience with leaf litter-based biogas plants', *Energy for Sustainable Development*, IX, 2, 49–62

Hemstock, S. L. (2005) *Biomass Energy Potential in Tuvalu* (Alofa Tuvalu), Government of Tuvalu Report

Hemstock, S. L. and Raddane, P. (2005) *Tuvalu Renewable Energy Study: Current Energy Use and Potential for Renewables* (Alofa Tuvalu, French Agency for Environment and Energy Management – ADEME), Government of Tuvalu

Matakiviti, A. and Kumar, S. D. (2003) Personal communication, South Pacific Applied Geoscience Commission and Fiji Forestry Department of Energy, Suva, Fiji Islands

Rosillo-Calle, F., Woods, J. and Hemstock, S. L. (2003) 'Biomass resource assessment, utilisation and management for six Pacific Island countries', ICCEPT/EPMG, Imperial College London, SOPAC – South Pacific Applied Geoscience Commission

Woods, J., Hemstock, S. L. and Bunyeat, W. (2005) 'Bioenergy systems at the community level in the South Pacific: impacts and monitoring, greenhouse gas emissions and abrupt climate change: positive options and robust policy', *Journal of Mitigation and Adaptation Strategies for Global Change* (in press)

CASE STUDY 7.4   BIODIESEL FROM COCONUT AND JATROPHA

Jeremy Woods and Alex Estrin

Introduction

Oil-bearing perennial plants are a common feature of the landscapes of tropical to semi-arid climates. Many of these plant types are little known to the world in general, although they are often widely used in traditional cultures. This is particularly the case from the prospective of producing biodiesel or pure plant oil which could, in some circumstances, be used for a diesel replacement fuel in transport and for stand-alone applications such as electricity generators. Two potential sources of oil for biodiesel production are assessed here but this certainly does not mean that many other promising candidates do not exist, some of which may have more beneficial yields and characteristics than those assessed below. In fact, according to Choo and Ma (2000), some of crops studied as potential biodiesel resources include: rape/canola, sunflower, coconut, maize, jatropha, cottonseed, peanut and palm. Of these crops, rape comprises more than 80 per cent of current global biodiesel production, sunflower 13 per cent, soy bean 1 per cent and palm oil 1 per cent. Other crops are at the R&D stage.

In this case study a brief methodology for assessing and calculating the energy production potential from the following two species is provided:

- *Cocos nucifera*, commonly known as the coconut;
- *Jatropha curcus*, often simply called 'Jatropha'.

Cocos nucifera – coconut

Coconut (*Cocos nucifera*) is grown in more than 93 countries and its production covers an area of 12 million ha producing over 10 million tonnes of copra (dried coconut flesh) equivalent. Coconut provides food, drink, medicine, shelter and is grown and relied on by millions of smallholders to provide their livelihoods. It has important social, cultural and religious connotations in many parts of the world and its products, including coconut oil, are consumed in more than 120 countries.

Historically, coconut activities have brought good returns from export earnings but recently the coconut has come under intense pressure from palm oil, mainly originating in the Far East. As a result, large areas of extensive coconut plantations, many of which exist on small tropical and sub-tropical islands, are being abandoned or becoming run down, with severe implications for the local populations who historically have depended on this plant for their livelihoods. Harvesting coconuts and using the oil-rich flesh and the energy-rich shells and residual plant material to produce energy may provide a new source of income for such, currently low value, plantations and their inhabitants. This option is evaluated for the island of Tuvalu in more detail below.

What is coconut?
Options for producing energy from coconuts include the use of:

- plant residues arising from the palm (trees) including the leaves, coconut husks and the main stem itself once past maturity;
- the fruits (known as 'nuts'; see Figure 7.1). The adult palm supports numerous bunches of nuts at all stages of maturity and which grow throughout the year. The nuts are protected by a thick fibrous husk and, once mature, fall to the ground without damage. Once on the ground, germination can occur over the next three to seven months, during which time the fruit can travel by sea currents over long distances to new locations. The nuts themselves contain a fibrous outer layer, a hard lignin-rich shell, soft oil-rich white flesh adhering to the inner surface of the shell in a relatively thin layer and an internal liquid or 'milk' which is mostly water but is nonetheless a nutritious drink.

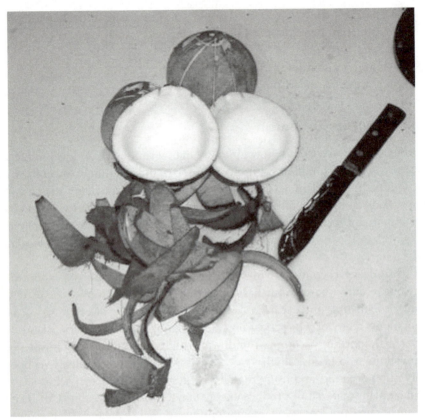

Figure 7.1 *Harvested coconuts and shell*

In coconut-derived biodiesel or pure plant oil (PPO) production for use as a diesel substitute, the flesh is the most important component of the nut and an average composition for the flesh is:

- un-dried:
  — 50 per cent water;
  — 34 per cent oil;
  — 16 per cent fats, proteins, etc.
- copra (dried flesh) is:
  — 5 per cent water;
  — 64 per cent oils.

The potential production of biodiesel or PPO depends on:

- the number of palms per unit land area (hectare);
- productivity of palms (nuts per palm);
- quality of nuts, their processing to copra and access to markets;

- efficiency of oil extraction from copra:
  — typically 400 g oil per kg copra;
  — recovers approximately 60 per cent of the total oil contained in the original copra.

Coconut oil is assumed to have an energy content of 43 GJ/t but it can vary between 37 and 43 GJ/t. The oil has a density of 0.91 kg/l which can be compared with mineral diesel which has an energy content of 46 GJ/t and a density of 0.84 kg/l.

The basic physical and energy properties of coconut oil are similar to mineral diesel. Coconut oil can be used directly in some diesel engines (after filtering and possibly pH adjustment) with modified fuel supply systems or in unmodified diesel engines if esterified.

Calculating oil and energy yields from coconut

A rough estimate of the potential oil and energy yields which can be obtained from coconut plantations can be calculated using the methodology and example outlined in Table 7.6.

Economics

The economics of biodiesel production from coconut will be highly site-specific, but in rural areas where coconut harvesting has become uneconomic for copra/oil exports and where imported fossil fuels and electricity have a high premium it may make economic sense to use coconut oil for mineral diesel substitution. In the case of Tuvalu, a strong macro-economic case exists for the use of copra-derived oil as outlined below. In other locations and times the economic basis may be very different for justifying coconut biodiesel production.

In late 2002, copra was being purchased by the Tuvalu Coconut Cooperative at A$1/kg (A$1000 per tonne), a price which also contained a social subsidy

Table 7.6 *Calculating potential oil and energy recovery rates from coconut*

	Low	Medium	High
Coconut trees per ha	151	254	351
Nuts per tree	50	80	120
Flesh per nut (kg; 50% moisture)	0.276	0.34	0.416
Shell per nut (kg)	–	0.2	–
Flesh per ha (kg; 50% moisture)	2084	7004	17522
Copra per nut (5% moisture)	1146	3852	9637
Recoverable oil per kg copra (kg)	0.3	0.4	0.55
Recoverable oil per ha (kg)	344	1541	5300
Recoverable litres oil/ha	378	1693	5825
Recoverable energy GJ/ha	14.8	66.3	227.9

Source: Woods and Hemstock (2003).

meant to enable copra production to continue on the outer islands. This price was equivalent to US$1.20 per litre coconut oil. At this time, the cost of imported diesel was US$0.55 per litre (2001/2002). However, the world market price for copra was about A$300–400/t copra (A$0.3–0.4/kg). In addition to the price paid by the cooperative for copra, it cost A$200 per tonne (A$0.2/kg) to ship the Copra to Fiji. Therefore, the total cost of delivering copra to Fiji was about A$1.3/kg compared to a world market price of about A$0.35/kg. Thus, Tuvalu was supporting copra production by about A$1 per kg or A$2 (US$1.20) per litre coconut oil.

If, instead of heavily subsidizing the export of copra, Tuvalu was to convert it to coconut-powered electricity production, Tuvalu would save itself US$0.55 per litre diesel (2002 prices), and US$0.12 per litre in copra shipping costs resulting in a total saving of US$0.67 per litre of coconut oil used for electricity production!

In effect, it is costing Tuvalu a minimum of US$0.78 to deliver coconut oil to Fiji but it is importing diesel at US$0.55 per litre.

By using coconut oil for electricity generation Tuvalu would save:

- the cost of imported diesel;
- the cost of subsidizing copra production and shipping.

However, additional costs would be incurred in using coconut oil for electricity production through the need to modify the fuel supply systems used in Tuvalu's diesel generators and in extracting the oil from the copra. On the other hand, the residual copra pulp would have value as pig feed. Other energy products could also be produced, including charcoal or producer gas from the coconut shells and husks. The husks also make a good soil amender for poor soils.

Jatropha curcas

Jatropha curcas is a perennial shrub of Latin American origin that is now widespread throughout arid and semi-arid tropical regions of the world. Although its origin lies in Central America, Portuguese seamen are believed to have taken some varieties of Jatropha to India from the Americas in the 16th century for its medicinal values. As a member of the Euphorbiaceae family, it is a drought-resistant perennial, living up to 50 years and it can be grown on marginal soils. It can grow into big trees, or as clusters of closely packed bushes. Whether it is a tree or a bush, it is always green and can be used as a living hedge for fencing in domesticated animals and for keeping wild animals out.

Depending on local climatic conditions, the trees can grow 7–10 m tall, but in rain-fed areas stunting occurs and the shrubs may only reach 2–3 m in height. In plantations, Jatropha is grown with a single trunk, two branches at each node and with the stem diameter varying between 250 and 100 mm. Jatropha can bear fruit for 25 years. The fruits are 25 mm long and oblong. In the beginning these are green, but as they ripen they turn yellowish with a golden tinge. On drying, they darken and when fully dry, are black. When the dry fruits are split, two seeds are found in each of the three pockets, with the seeds comprising about 65 per cent of the total mass of the dry fruits. The seeds are oil rich and when extracted can be used for soap, oil and/or biodiesel making.

The seeds contain:

- 19.0 per cent oil;
- 4.7 per cent polyphenol.

The energy content (HHV) for:

- the seed (0 per cent moisture content) is 20 MJ/kg;
- the oil – 37.8 MJ/kg.

The oil fraction consists of:

- saturated fatty acids, palmitic acid (14.1 per cent), stearic acid (6.7 per cent), and
- unsaturated fatty acids, oleic acid (47.0 per cent), and linoleic acid (31.6 per cent).

Jatropha can be planted at spacing of between 0.2–2.5 m depending on the soil type and can survive for 30 to 50 years. Its cetane number is 51 while that of diesel is 45, indicating that Jatropha oil does not ignite as readily as mineral diesel. However, a major disadvantage of raw Jatropha oil is that it solidifies at higher temperatures than mineral diesel thereby obstructing the flow to the burner or the engine when air temperatures fall.

Jatropha can be grown successfully in most categories of cultivatable lands except waterlogged lands and marshes and deserts. Most parts of tropical and sub-tropical areas are ideal for Jatropha plantation. It can grow in soft, rocky, sloping soils along a mountain as well as in medium fertile lands. It can be grown along canals, water streams, boundaries of crop fields, along the roads, along railway lines. The pH of soil should be 5.5 to 6.5 and its minimum rainfall threshold is 500–750 mm.

Jatropha can be propagated by sowing seeds or by directly planting short stem sections of existing stems or branches. Fertilizer applications of 50 kg of urea, 300 kg of single super phosphate and 40 kg of potassium nitrate per ha are advised to be applied annually.

Oil and energy yields from Jatropha
Planting densities and yields are highly site-specific and the development status of Jatropha as a commercial crop is at a very early stage, meaning that great care

should be taken in evaluating potential yields that could be obtained. An example calculation is provided here for reference:

- Assuming 2500 plants per hectare can be planted and from each plant at least 1 kg of Jatropha fruit is obtained annually,[1] then the total fruit mass obtained per hectare will be 2500 kg.
- The seed mass will be 1625 kg (65 per cent of the fruit mass).
- Assuming a 60 per cent oil content of the seeds, the total amount of oil that can be recovered from Jatropha fruit will be 975 kg per hectare, i.e. 36,855 MJ per hectare.
- The extraction efficiency of oil expellers is approximately 60 per cent and so the final recoverable oil yield is calculated as 22,000 MJ per ha (585 kg oil).

The seeds are 10–20 mm long and each weighs 0.5–0.7 g. Jatropha seeds have the following average composition (dry mass):

- 6.2 per cent moisture;
- 18 per cent protein;
- 38 per cent fats;
- 17 per cent carbohydrates;
- 15 per cent fibre;
- 5.3 per cent ash.

Oil can be expelled from seeds in normal screw-type engine-driven oil expellers, by manual Bielenberg-ram-presses or from the sediment of the oil purification process. The expeller crushes seeds along with the shell. The oil is then immediately filtered through a filter press and used for the manufacture of biodiesel.

Typically, in hedgerow cultivation the production of seeds is estimated to be about 0.8 kg per metre of hedge per year (Henning, 2000). The beans of Jatropha contain about 35 per cent viscous, non-edible oil, which can be used for the production of a raw material for cosmetic products, as fuel for cooking and lighting and as a substitute for diesel fuel (Bhattacharya and Joshi, 2003). The high quality oil may be also used for soap making in rural areas, giving local women the chance to gain additional income and thus strengthen their economic position. The press-cake, another extraction by-product, can be used as a high-grade organic fertilizer or sent to biogas plants (Bhattacharya and Joshi, 2003).

Note

1   This seems too low; something between 1–5 kg per plant can be expected.

References and further reading

Bhattacharya, P. and Joshi, B. (2003) 'Strategies and institutional mechanisms for large scale cultivation of *Jatropha curcas* under agroforestry in the context of the proposed biofuel policy of India', Indian Institute of Forest Management, *ENVIS Bulletin on Grassland Ecosystems and Agroforestry*, vol 1, no 2, Bhopal, India, pp58–72

Choo, Y.-M. and Ma, A.-N. (2000) 'Plant Power', *Chemistry & Industry*, 16, 530–534

ECOPORT/FAO (2006a) '*Cocos nucifera* "Coconut" ', http://ecoport.org/ep?Plant=744 (accessed 16 February 2006)

ECOPORT/FAO (2006b) '*Jatropha curcus*', http://ecoport.org/ep?Plant=1297 (accessed 16 February 2006)

Henning, R. K. (2000) 'Use of *Jatropha curcas* oil as raw material and fuel: an integrated approach to create income and supply energy for rural development; experiences of the Jatropha Project in Mali, West Africa', Weissenberg, Germany, available at www.jatropha.org (accessed 17 February 2006)

Lele, S. (2004) *Biodiesel in India*, 44pp, Vashi, Navi Mumbai, India, available at http://www.svlele.com/biodiesel_in_india.htm (accessed 17 February 2006) and/or http://business.vsnl.com/nelcon/intro.htm

Woods, J. and Hemstock, S. (2003) 'Tuvalu coconut oil bioelectricity potential' http://www.iccept.imperial.ac.uk/research/projects/SOPAC/index.html (accessed 16 February 2006)

CASE STUDY 7.5   THE ROLE OF BIOMASS IN CARBON STORAGE AND CLIMATE CHANGE

Peter Read[1]

Introduction

Bioenergy is, from the perspective of climate change and the carbon cycle, scientifically different from other zero-emissions (and mostly renewable) energy systems. This is because its initial stage, the production of biomass raw material, actively removes CO_2 from the atmosphere through the process of photosynthesis.

The other zero-emissions systems simply avoid adding to the stock that is already there. As a consequence, even the universal adoption of those systems could achieve no more than asymptotic progress towards the level of CO_2 in the sinks into which atmospheric CO_2 spills. These are the biosphere (liable to reverse and become a net emitter under temperature stresses foreseeable with business-as-usual emissions scenarios (Cox et al, 2006)) and the ocean surface layers, where it forms carbonic acid (already at a concentration that is threatening the food chain for ocean ecosystems (Turley et al, 2006)) which, in turn, spill into the deeper ocean.

During the late 1990s, the concept of CO_2 capture, compression and sequestration (CCS) came to the fore as a technology for turning the use of coal at large point sources of CO_2 emissions, typically thermal power stations, into a zero-emissions energy system (or near-zero if the capture technology is less than 100 per cent efficient (IPCC, 2005)). Shortly afterwards the notion of linking bioenergy to CO_2 sequestration (BECS) was announced (Obersteiner et al, 2001).

BECS constitutes a negative emissions energy system in which the more bioenergy products are consumed, the less CO_2 remains in the atmosphere. By using revenues from bioenergy product sales to pay for both the acquisition of biomass raw material and the safe disposal of CO_2 waste products, BECS actively pumps CO_2 out of the atmosphere. On a sufficiently large scale, this can result in a reduction in atmospheric CO_2 levels to below the asymptotic path mentioned above.

Addressing potential abrupt climate change – an expert workshop

It was noted early in the debate (Schelling, 1992) that 'insurance against catastrophes is an argument for doing something expensive about greenhouse gas emissions'. This applies even in developed countries, where high incomes are largely derived from activities that are insulated from the impacts of gradual climate change. Also, Article 3.3 of the UNFCCC places an obligation on the parties to take such precautionary action.

A decade later, concern has risen that abrupt climate change events are inevitable (Alley et al, 2001) although the National Academy of Science authors did not claim that such events are necessarily in the high damage category. But more recently the Stabilisation2005 Symposium (Schellnhuber et al, 2006) drew attention to research that suggests the climate system may be near a threshold for such a high damage event, and subsequent media reports cover happenings that could be regarded as precursors for such change. These include a measured slowing of the 'Gulf Stream', the release of methane from thawing West Siberian tundra and the retreat of the southern summer limit of Arctic sea ice cover.

Against this background, and the new awareness of the potential of BECS, the Better World Fund of the UN Foundation supported an Expert Workshop that was held in Paris at the end of September 2004. This had a mission statement 'to address the policy implications of potential abrupt climate change' (visit www.accstrategy.org for details, and the link to /simiti for subsequent peer-reviewed papers, forthcoming in a special issue of *Mitigation and Adaptation Strategies for Global Change* Vol 11, No 1).

The conclusion reached by the workshop was that 'Policymakers should be urged to stimulate the growth of a global bioenergy market, with world trade (mainly 'South–North' trade) in liquid biofuels such as ethanol and synthetic (e.g. Fischer Tropsch) biodiesel' (Read, 2006a).

This was seen as the first stage in a two-stage strategy (Read and Lermit, 2005) to enable the global community to be prepared for scientific evidence that potential climate change had become imminent abrupt climate change. It would be low cost, possibly negative cost, depending on the future evolution of world oil prices, but would have a long time-frame. This is due to the need to initiate a very large number of sustainable land improvement projects worldwide to enable the co-production of biomass for energy along with traditional outputs from farming and forestry.

The second stage – linkage to CCS to comprise BECS – would possibly be costly (although justifiable in circumstances of imminent abrupt climate change) but could be accomplished relatively quickly providing the necessary ground-work had been carried out during implementation of the first stage. Such groundwork would include the design of bioenergy systems for subsequent retro-fitting with CCS, and prospecting for appropriate CO_2 disposal sites such as deep saline aquifers.

Storage versus sequestration

The word 'storage' is used in this case study to connote a broader meaning than 'sequestration', which, particularly in North America, has come to be associated with CCS and, in some other contexts, with permanent new forest plantations.[2] The broader meaning arises as an outcome of the increased knowledge base provided by the expert workshop and covers a wider variety of final destinations for carbon extracted from the atmosphere by photosynthesis.

These include the enhanced quantity of labile carbon that can result from increased soil productivity, both in-soil and in leafy material, along with specific managed increases of long-lived in-soil carbon. Among the latter are the use of 'ramial' wood from deciduous trees to promote the white fungal activity that provides the basis for soil formation and deep accumulation in temperate zones (Caron et al, 1998) and the use of biochar[3] as a soil amendment agent. Known in Brazil as 'terra preta' (Portuguese for 'black earth') this technology provided the basis for pre-Columbian agriculture on the otherwise infertile yellow soils of the Amazon basin and is the origin of the 'El Dorado' myth.[4]

Biochar soil amendment is also a traditional land improvement technique in Japan and there are currently projects to demonstrate its potential in Indonesia (under the Clean Development Mechanism (CDM)), in Australia (as Joint Implementation) and in Japan (for multiple domestic environmental objectives) (Ogawa et al, 2006). Estimates of its global mitigation potential suggest that terra preta alone could absorb several decades of energy sector carbon emissions (Lehmann et al, 2006). It is a technique that could prove particularly relevant in parts of sub-Saharan Africa where the geological prospectivity of deep saline aquifers is poor (Haszeldine, 2006) and where increased soil productivity could provide an important dimension of sustainable rural development.

Holistic mitigation strategy

The availability of diverse potential storages for carbon taken from the atmosphere has led to an appreciation that the notion of reducing anthropogenic CO_2 emissions, that is embodied in the Kyoto Protocol, may not be the best way to mitigate climate change, whether gradual or potentially abrupt. The economic theory on which that approach is based treats the problem of reducing pollution – say, of sewage into a river – to an acceptable level – say, that precludes excessive fish kill. However, the terrestrial biosphere emits and absorbs twenty times as much of the same material, CO_2, as is anthropogenically emitted.

Thus, anthropogenic CO_2 emissions are not a flow pollution problem in the sense that has been analysed into economic theory. The greenhouse gas mitigation problem is analogous to adding more pure water to the river so as to increase its total flow into a lake, sufficiently to raise the stock of water in that lake and create the danger of flooding a lakeside city. In that situation it is sensible to look at all the flows of water into the stream, and also at how water gets out of the lake, and where it gets to afterwards. That is the approach taken in the holistic mitigation strategy (Read and Parshotam, 2006) which, coincidentally, offers not only more effective mitigation than can be envisaged under a Kyoto-style emissions cap, but also numerous co-lateral benefits to key geo-political interests, as outlined elsewhere (Read, 2006b).

Figure 7.2 illustrates the effect of this approach through three illustrative technologies involving both bioenergy production and carbon storage. These are ~400 million ha sugar cane, with fermentation to ethanol and ligneous residues used for power generation; ~700 million ha of switchgrass, with extraction of protein for animal feed, fermentation of cellulosic fractions to ethanol and power generation from residues; and 1 billion ha of 25-year rotation forest and co-production of timber with ethanol and power. Storages of carbon arise from the standing timber, from storage of CO_2 of fermentation at low cost and higher cost

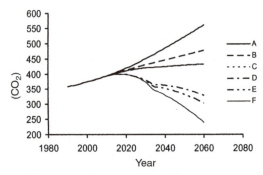

Source: Read and Parshotam, 2006, with permission.

Figure 7.2 *Bioenergy with carbon storage (BECS) – impacts on level of CO_2 in the atmosphere*

CCS applied to flue gases from biomass-fuelled thermal power generation, along with the carbon stored in fossil fuel that is left underground, displaced by biofuels.

Line A in Figure 7.2 replicates the path of carbon in atmosphere under the high material growth, consumption-oriented, SRES A2 climate change scenario of the International Panel on Climate Change (IPCC). This falls to line B with the sugar cane activity alone and to line C with both sugar cane and switchgrass, in each case with no CO_2 storage. Line D shows the outcome with the forestry activity also added. It should be noted that the bulk of the early impact prior to 2025 (when no further areas of plantation are added and continued planting is on the land cleared by last year's crop – see Note 2) is due to the increasing stock of carbon held up in standing timber as 40 million ha are planted each year for 25 years and as each preceding year's plantings continue to grow more biomass, until maturity and eventual felling.

Figure 7.3 illustrates the consequences of expanding the global bioenergy market without regard for sustainability issues. Lines A and D are as before but lines G, H and I demonstrate the effect on line D if the land use changes involved result in the release of 30, 90 and 300 tons of carbon per hectare through disturbance of in-soil carbon and burn-off to clear existing vegetation (with the latter broadly corresponding to the destruction of dense tropical rain-forest to make way for palm oil plantations, as is reportedly happening in Indonesia (Monbiot, 2005)).

Thus, the vision that the 'polluter pays' principle can be turned to a greening of the earth, and to the advancement of those developing countries located in regions of high potential net primary soil productivity, needs to be tempered with an awareness of the dangers inherent in a global bioenergy programme driven simply by concerns for energy security. If the climate change problem turns out to be relatively short term, due to imminent abrupt climate change, such a programme may do more harm than good.

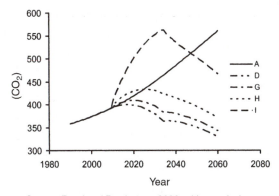

Source: Read and Parshotam, 2006, with permission.

Figure 7.3 *Bioenergy with CO_2 release at time of land use change*

Even though experience with well-intentioned land use change has not always been happy (Woods et al, 2006) the assumption in the holistic strategy is that ongoing commercial demands for food, fibre and fuel, conditioned by climate change policy measures, and mediated by monitoring and certification procedures, will succeed where good intentions have sometimes failed. In particular, participation by selected developing countries in such a programme, to the mutual prospective benefit of exporting countries' sustainable development and importing countries' energy security and climate change mitigation commitments, must be conditional on the exporting countries' agreeing to externally negotiated and transparently monitored and certified sustainable development criteria.

Success will be measured by reconciling worldwide market demands for these traditional goods, and for the new public good of climate change mitigation, with local socio-economic development and environmental quality. For this, a participatory framework that ensures the involvement and commitment of populations living on the land in question is needed. So an essential first step, as regards the developmental potential of the holistic strategy, is a large-scale capacity building programme to train people in the countries involved to initiate country-driven projects that align the aspirations of local communities with the national economic growth strategy, and to provide ongoing technical and commercial back-up.

Subject to land use controls to shape the pattern of land use change, there is, under the holistic strategy, *prima facie*, enough land left over from commercial uses to provide the conservation areas, migration routes, etc., that are needed to sustain remaining global biodiversity. This may be funded out of the wealth created by policy-driven investment in land-using technologies and, maybe, also by eco-tourism.

Conclusion

Owing to specific scientific properties, not shared with other renewable energy technology, biomass, as raw material for a global bioenergy market, has a key role to play in any climate change mitigation programme. Linked to a variety of carbon storages, it has a unique role in the event that abrupt climate change is proven to be imminent. Thus, the development of such a market provides the unique action that is available in response to Article 3.3 of the UNFCCC. The G8 commitment to a Global Bioenergy Partnership, and its linked commitment to capacity building to enable developing country participation in that Partnership, could provide the springboard for that response. This is providing the Partners look outside their domestic borders to meet their mitigation commitments, going beyond technology sharing to a willingness to rely, in part, on biofuel imports to meet domestic programmes for biofuel expansion. However, the risks to climate change mitigation, and to sustainable development from that Partnership, if pursued solely with a view to energy security, are manifest.

Notes

1   Of Massey University, New Zealand.
2   This concept has been subject to sharp criticism from environmentalist groups who argue that there can be no guarantee of permanence to forests that are at risk from fire, pests and other natural hazards. However, in the context of commercial forestry, where a 'normal' forest is always on average half grown, with equal area stands of every cohort from newly planted seedlings to mature trees due for felling (for co-produced timber and fuel) and replanting, there is a permanent storage of carbon for as long as the commercial incentive remains (i.e. in the present context, for as long as concerns over climate change and demand for forest products last).
3   The product of biomass pyrolysis, widely misnamed 'charcoal' – see Read, 2006a.
4   No Europeans revisited the Amazon for several decades after the first explorers, and when they did the pre-Columbian civilization had disappeared, with the large populations wiped out by measles and smallpox left behind by the first explorers.

References

Alley, R. B. et al (2001) *Abrupt Climate Change: Inevitable Surprises*, National Academy of Science Report, National Academy of Science Press, Washington, DC

Caron, C., Lemieux, G. and Lachance, L. (1998) 'Regenerating soils with ramial chipped wood', Publication no 83, Department of Wood and Forestry Science, Laval University, Quebec (http://www.sbf.ulaval.ca/brf/regenerating_soils_98.html)

Cox, P., Huntingford, C. and Jones, C. D. (2006) 'Conditions for sink-to-source transitions and runaway feedbacks from the land carbon cycle', Chapter 15 in Schellnhuber, H. J., Cramer, W., Nakicenovic, N., Wigley, T. and Yohe, G. (eds), *Avoiding Dangerous Climate Change*, Cambridge University Press, Cambridge, pp155–162

Haszeldine, R. S. (2006) 'Deep geological carbon dioxide storage: principles, and prospecting for bio-energy disposal sites', Article 3 in 'Addressing the Policy Implications of Potential Abrupt Climate Change: A Leading Role for Bio-Energy', *Mitigation and Adaptation Strategies for Global Change*, vol 11, no 1, pp377–401

IPCC (2005) *Special Report on Carbon Dioxide Capture and Storage*, Cambridge University Press, Cambridge

Lehmann J., Gaunt, J. and Rondon, M. (2006) 'Bio-char sequestration in terrestrial ecosystems – a review', Article 4 in 'Addressing the Policy Implications of Potential Abrupt Climate Change: A Leading Role for Bio-Energy', *Mitigation and Adaptation Strategies for Global Change*, vol 11, no 1, pp403–427

Monbiot, G. (2005) 'The most destructive crop on earth is no solution to the energy crisis', *Guardian*, 6 July 2005, London, p17

Obersteiner, M., Azar, C. Kauppi, P., Mollerstern, M., Moreira, J., Nilsson, S., Read, P., Riahi, K., Schlamadinger, B., Yamagata, Y., Yan, J. and van Ypersele, J.-P. (2001) 'Managing climate risk', *Science*, 294, (5543), 786b

Ogawa, M., Okimori, Y. and Takahashi, F. (2006) 'Carbon sequestration by carbonisation of biomass and forestation: three case studies', Article 5 in 'Addressing the Policy Implications of Potential Abrupt Climate Change: A Leading Role for Bio-Energy', *Mitigation and Adaptation Strategies for Global Change*, vol 11, no 1, pp429–444

Read, P. (ed.) (2006a) 'Addressing the Policy Implications of Potential Abrupt Climate Change: A Leading Role for Bio-Energy', *Mitigation and Adaptation Strategies for Global Change*, vol 11, no 1, pp501–519

Read, P. (2006b) 'Clearing away carbon', *Our Planet*, 16/4, (Special issue on Renewable Energy), pp28–29

Read, P. and Lermit, J. (2005) 'Bio-energy with carbon storage (BECS): a sequential decision approach to the threat of abrupt climate change', *Energy*, 30, 2654–2671

Read, P. and Parshotam, A. (2006) 'Holistic greenhouse gas management', *Climatic Change*, under review

Schelling, T. C. (1992) 'Some economics of global warming', (Presidential Address) *American Economic Review*, 82/1, 1–14

Schellnhuber, H. J., Cramer, W., Nakicenovic, N., Wigley, T. and Yohe, J. (eds) (2006) *Avoiding Dangerous Climate Change*, Cambridge University Press, Cambridge

Turley, C., Blackford, J. C., Widdicombe, S., Lowe, D., Nightingale, P. D. and Rees, A. P. (2006) 'Reviewing the impact of increased atmospheric CO_2 on ocean pH and the marine ecosystem', Chapter 8 in Schellnhuber, H. J., Cramer, W., Nakicenovic, N., Wigley, T. and Yohe, G. (eds), *Avoiding Dangerous Climate Change*, Cambridge University Press, Cambridge, pp65–70

Woods, J., Hemstock, S. and Burnyeat, W. (2006) 'Bio-energy systems at the community level in the South Pacific: impacts and monitoring', Article 7 in 'Addressing the Policy Implications of Potential Abrupt Climate Change: A Leading Role for Bio-Energy', *Mitigation and Adaptation Strategies for Global Change*, vol 11, no 1, pp461–492

Glossary of Terms

Accessible fuelwood (woodfuel) supplies   Are the quantities of wood that can actually be used for energy purposes under normal conditions of supply and demand

Agrofuels   Are biofuels obtained either as a product or by-product of agriculture; the term covers mainly biomass materials derived directly from *fuel crops* and *agricultural, agro-industrial and animal by-products* (FAO)

Air-dry   The term refers to the stage after the fuel has been exposed for some time to local atmospheric conditions, between harvesting and conversion of the fuel either to another fuel or by combustion to heat energy

Air-dried weight   This is the weight of wood in air-dry state, after being exposed over time to local atmospheric conditions. Weight of wood can be given in air-dry state or wet state. It may contain between 8 and 12 per cent (dry basis) moisture

Air-dried yield (wood)   The approximate mass of air-dry wood which would be obtained on drying or wetting of a sample of wood per unit volume of a sample

Air-dry density   The density based on the weight and volume of wood in equilibrium with the atmospheric conditions

Alcohol fuels   A general term which denotes mainly ethanol, methanol and butanol, usually obtained by fermentation, when used as a fuel

Ancillary energy   That energy required to produce inputs to agricultural production consumed in a single activity or production period; e.g. the energy sequestered in fertilizers, chemicals, etc.

Animal waste   The dung, faeces, slurry or manure which is used as the raw material for a biogas digester

Animate energy   Work performed by animals and humans. This is a very important source of power in many developing countries for agriculture and small-scale industry. Animals and humans provide pack and draught power, and transport by bicycle, boat, etc. (See Appendix 4.3 Measuring animal draught power.)

Ash content   The weight of ash expressed as a percentage of the weight before burning of a fuel sample burned under standard conditions in a laboratory furnace. The higher the ash content, the lower the energy value of the fuel

Bagasse   The fibrous residue from sugarcane which remains after the juice has been extracted. It constitutes about 50 per cent of cane stalk by weight and with a moisture content of 50 per cent its calorific value varies from 6.4 to 8.60 GJ/t. It is widely used to generate electricity and also as animal feed, in ethanol production, for pulp and paper, paperboard, furniture, etc.

Bark   A general term for all the tissues outside the cambium in sterns of trees; the outer part may be dead, the inner part is living

Basal area   The cross-sectional area of a tree estimated at breast height; it is normally expressed in m^2. The sum of the basal areas of trees (G) on an area of one ha is symbolized by $G = m^2/ha$. (Basal area is usually measured over bark.) Common values in young plantations are between 10–20 m^2/ha rising to a maximum of around 60 m^2/ha in exceptional circumstances in older plantations

Biofuels   A general term which includes any solid, liquid or gaseous fuels produced from organic matter, either directly from plants or indirectly from industrial, commercial, domestic or agricultural wastes (see also under **Bioenergy**)

Biogas   The fuel produced following the microbial decomposition of organic matter in the absence of oxygen. It consists of a gaseous mixture of methane and carbon dioxide in an approximate volumetric ratio of 2:1. In this state the biogas has a calorific value of about 20–25 MJ/m^3 but this can be upgraded by removing the carbon dioxide

Biomass   The organic material derived from biological systems. Biological solar energy conversion via the process of photosynthesis produces energy in the form of plant biomass which is about ten times the world's annual use of energy. Biomass does not include fossil fuels, although they also originate from biomass-based sources. For convenience, biomass might be sub-divided into two main categories: **woody biomass** and **non-woody biomass** (see under these terms) although there is no clear cut distinction between these terms

Biomass conversion process   The methods which convert biomass into fuel can be classified as:

1   biochemical, which includes fermentation and anaerobic digestion;
2   thermo-chemical, which includes pyrolysis, gasification and liquefaction. See under these terms

Bioenergy or **biomass energy**   Covers all energy forms derived from organic fuels (biofuels) of biological origin used for energy production. It comprises purpose-grown energy crops, as well as multipurpose plantations and by-products (residues and wastes). The term 'by-products' includes solid, liquid and gaseous by-products derived from human activities. Biomass may be considered as one form of transformed solar energy (FAO). There are two main types of biomass energy: **biomass energy potential** and **biomass energy supply** (see under these terms)

Biomass energy potential   This term refers to the total biomass energy generated per annum. This represents all the energy from crop residues, animal wastes, the harvestable fuel crops and the annual increase in the volume of wood in the forests

Biomass energy supply   The term represents the total biomass that is accessible to the market taking into account the logistics of collecting from various sources

Biomass inventory   All living organisms and dead organic material in the soil (humus) and in the sea. A total of 99 per cent of living organic material is plant biomass, produced mainly by forests, woodland, savannas and grassland

Biomass productivity   The increase in wet or dry weight of living and dead plant material/unit area or individual/unit time; biomass productivity may be expressed, for example, on the basis of the whole tree or parts thereof

Bole   Primary length of a tree; the trunk of a tree

Bone dry   See **oven-dried**

Breast height diameter (commonly known as **dbh**, see under **Diameter at breast height**)

Brix   The Brix Scale is a measure of sugar in a solution. Computed in 'degrees Brix' (°Bx), it is the percentage of pure sucrose to water at a particular temperature, either by volume ('volume Brix') or by weight ('weight Brix'). A 15 °Bx solution therefore has 15 gm of sugar per 100 gm of water. See also **pol**

Brushwood   Shrubby vegetation and stands of tree species that do not produce commercial timber

Bulk density   The mass (weight) of a material divided by the actual volume it displaces as a whole substance expressed in lb/ft^3, kg/m^3, etc.

Burning index   A number in an arithmetical scale, determined from fuel moisture content, wind speed and other selected factors that affect burning conditions and from which ease of ignition of fires and their probable behaviour may be estimated

Calorific value   A measure of the energy content of a substance determined by the quantity of the heat given off when a unit weight of the substance is completely burned. It can be measured in calories or joules; the calorific value is normally expressed as kilocalories/kg or MJ/kg

Canopy   The total leaf cover produced by branches of forest trees and leaves of other plants, affording a cover of foliage over the soil

Charcoal   The residue of solid non-agglomerating organic matter, of vegetable or animal origin, that results from carbonization by heat in the absence of air at a temperature above 300 °C

Combustion energy   The energy released by combustion. It is normally taken to mean the energy released when substances react with oxygen. This may be very fast, as in burning, or slow, as in aerobic digestion. Technical temperature achieved depends on whether combustion is with air or pure oxygen, and on the method of combustion

Commercial forestry   The practice of forestry with the object of producing timber and other forest produce as a business enterprise

Conventional energy (or fuel)   Denotes energy sources which have hitherto provided the bulk of the requirements of modern industrial society (e.g. petroleum, coal, and natural gas; wood is excluded from this category). The term is almost synonymous with *commercial energy*

Conversion efficiency   The percentage of total thermal energy that is actually converted into electricity by an electric generating plant

Cookstove   Widely used for cooking food in developing nations. There are many types; they use mostly biomass fuels, particularly wood. Stoves have very low thermal efficiency – about 13–18 per cent

Cord   A measure of stacked wood. By definition one cord is a stack of pieces of wood which is 4 ft wide, 4 ft high and 8 ft long giving a total of 128 ft^3 (0.00384 m^3). In practice the weight and volume in a cord can vary appreciably

Crop residue index (CRI)   A method used for estimating crop residues. This is defined as the ratio of the residue produced to the total primary crop produced for a particular species. The biomass produced by crop plants is usually one to three times the weight of the actual crop itself. The CRI is determined in the field for each crop and crop variety, and for each agro-ecological region under consideration

Current annual increment (CAI)   The increment of total biomass produced over a period of one year. It must be distinguished between increment of the standing crop and net production. The difference is the amount lost by litter fall, root slouching and grazing

Decentralized energy   The energy supplies generated in dispersed locations and used locally, maintaining a low energy flux from generation to supply. The term is often used for renewable energy supplies since these harness the energy flows of the natural environment, which are predominantly dispersed and have relatively low energy flux. In contrast, the centralized energy supplies of large-scale fossil and nuclear sources produce large energy fluxes which are most economically used in a concentrated manner

Delivered energy   The actual amount of energy available or consumed at point of use. The concept recognizes that in order to have a unit of economically usable energy, there are prior exploration, production and delivery systems, each of which detracts from the next amount of energy delivered to the point of use. It is also called *received energy* since it records the energy delivered or received by the final consumer

Densified biomass fuels   Fuel made by compressing biomass to increase the density and to form the fuel into a specific shape such as briquettes, pellets, etc. to ease transportation and burning

Density   The weight of unit volume of a substance. In the case of wood several different densities can be referred to:

- basic density (the weight of dry matter in unit volume of freshly felled wood);
- air-dry density (the weight of unit volume of oven-dried wood);
- stacked density (the weight of wood at stated moisture content – fresh, air dry, etc. – contained in a stack of unit volume).

Diameter at breast height   Method used by foresters to determine total height and crown measurements (diameter + depth) to estimate individual tree volumes, mean height, basal area at breast height and mean crown measurements. The reference diameter of the main stem of standing trees is usually measured at 1–3 m above the ground level. This allows the tree crop volume per unit area to be estimated

Direct combustion   Complete thermal breakdown of organic material in the presence of air so that all its energy content is released as heat

Dry basis   A basis for calculating and reporting the analysis of a fuel after the moisture content has been subtracted from the total. For example, if a fuel sample contains A% ash and M% moisture, the ash content on a dry basis is: $100:100 - A\%$

Dry fuel   Biomass materials with low moisture content, generally 8–10 per cent. The allowable moisture content for dry fuel varies with requirements of the combustion system

Dry ton   Biomass dried to a constant weight of 2000 pounds

Energy content   The intrinsic energy of a substance, whether gas, liquid or solid, in an environment of a given pressure and temperature with respect to a data set of conditions. Any change of the environment can create a change of the state of the substance with a resulting change in the energy content. Such a concept is essential for the purpose of calculations involving the use of heat to do work. See also **heating value**

Energy content as received   The energy content of a fuel just before combustion. It reflects moisture content losses due to air drying or processing. For this reason, the energy content as received is generally higher per unit weight than that of the fuel at harvest

Energy content of fuel at harvest   Normally used for biomass resources, it refers to the energy content of a fuel at the time of harvest. It is also referred to as the **green** energy content

Energy efficiency   The percentage of the total energy input that does useful work and is not converted into low-quality, essentially useless, low temperature heat in an energy conversion or process

Ethanol fuel (bioethanol)   Fermentation ethanol obtained from biomass-derived sources (usually sugar cane, maize, etc.) used as a fuel. Ethanol is also obtained from fossil fuel (e.g. coal and natural gas)

Fermentable sugars   Sugar (usually glucose) derived from starch and cellulose that can be converted to ethanol. Also known as reducing sugars or monosaccharide

Firewood   See **woodfuels** and **fuelwood**

Forest inventory   Forests have different structures, that is the different species, ages and sites of trees are grouped in different patterns in different forests. An inventory of a forest area can provide information for many different purposes, e.g. it may be part of a natural resource survey, a national project to assess the potential of forest, etc. Forest inventory thus refers techniques used for measuring of trees

Forest mensuration   Branch of forestry concerned with the determination of the dimensions, form, increment and age of trees, individually or collectively, and of the dimensions of their products particularly sawn timber and logs

Fuelwood (or **firewood**)   See **woodfuels**. Fuelwood is the term used by FAO

Fuelwood needs   Minimum amount of fuelwood/woodfuel necessary in view of the minimum energy estimated to be indispensable for household consumption, artisanal purposes and rural industries, in line with local conditions and the share of fuelwood in their energy supplies

Fuelwood supply   Is the quantity of fuelwood/woodfuel which is, or can be, made available for energy use on the basis of the mean annual productivity of all potential resources on a sustainable basis

Green fuel   A term that denotes freshly harvested biomass not substantially dehydrated, with varying moisture content, obtained from processed vegetal wastes and organic material. It also denotes biomass-based fuels

Green manure   Fresh or still-growing green vegetation ploughed into the soil to increase the organic matter and humus available to support crop growth

Green weight   The weight of freshly cut wood which contains 30–35 per cent (dry-basis) moisture

Gross heating value (GHV)   This is a measure of the total heat energy content of a fuel, and equals the heat released by the complete combustion of a quantity of fuel. It is also referred to as the *higher heating value*

Gross increment   The total increase in growth of trees of all diameters down to a stated minimum diameter over given period. Gross increment includes both the increment in the growth of trees that fulfil the diameter criterion and the trees that reach the minimum diameter after this period

Gross primary production (Pg) (biomass)   The total amount of photo-synthetically produced organic matter assimilated by a community or species per unit land area per unit time

Growing stock   Refers to the living part of the standing volume of trees (see **standing volume**)

Harvest index   The index of a given crop is obtained by dividing the weight of the useful harvested part by the total weight of organic material produced by the crop (see also Appendix V)

Heating value   The heating value (HV) of fuels is recorded using two different types of energy content:

- **gross heating value (GHV)**
- **net heating value (NHV)**

(see under these terms). Although for petroleum the difference between the two is rarely more than 10 per cent, for biomass fuels with widely varying moisture contents, the difference can be quite large. Heating values of biomass fuels are often given as an energy content per unit weight or volume at various stages:

green, air-dried and **oven-dried** material (see under these terms). See also **energy content**

Leaf area index (LAI)   The leaf surface area (one side only) in a stand of a given crop covering a unit area of land surface. Optimum photosynthesis conditions are achieved

Lower heating value (LHV)   See **net heating value**

Maximum sustainable yield   The largest amount which can be taken from a renewable natural-resource stock per period of time, without reducing the stock of the resource

Mean annual increment (MAI)   Average annual increment over a given number of years; it is obtained by dividing the total biomass produced by the number of years taken to produce it

Measurement units   There are four basic types of units used in energy measurements:

- **stock energy units**;
- flow or rate energy units;
- specific energy consumption or energy intensity;
- **energy content** or **heating value**;

(see further information under the bold terms)

Moisture content   The moisture content is the amount of water contained in a fuel. For biomass fuels special care must be taken to measure and record the moisture content. The moisture content can change by a factor of 4–5 between initial harvesting and final use and is critical both to the heating valve on a weight or volume basis and to differences between GHV and NHV. See Appendix V for further details

Moisture content dry basis (mcwb)   The ratio of the weight of water in the fuel to the oven-dried (solid fuel) weight, expressed as a percentage

Moisture content wet basis (mcdb)   Refers to the ratio of the weight of water in the fuel to the total weight of fuel. The mcwb is expressed on a percentage basis

Net energy ratio (NER)   The NER of a given biomass fuel process may be calculated by dividing the heat content of the final fuel product by the energy (heat) equivalent of all the processes and machinery that were used in the process

Net heating value (NHV)   The potential energy available in the fuel as received, taking into account the energy that will be lost in evaporating and superheating the water when the woodfuel burns. This energy loss accounts for most of the reduction in efficiency when systems are fuelled with green wood rather than dry wood or fossil fuels. It is sometimes referred to as the **lower heating value**

Net increment   Average net increment *less* natural losses over a given period

Net primary productivity   Refers to the amount of organic matter formed by photosynthesis per unit of time, i.e. the amount remaining in the plant after respiration. Usually expressed in terms of dry weight

Net primary production (Pn)   The total amount of photosynthetically produced organic matter assimilated less that lost due to plant respiration, i.e. the total production which is available to other consumers. Primary production is the production of vegetation and other autotrophic organisms. Pn denotes a gain of material by a plant community; it can be determined from the sum of changes in plant biomass together with losses of plant material (e.g. death, etc.) over a given time interval. Pn is measured typically as grams per unit area or volume See also **primary productivity**

Non-woody biomass   There is no clear division between this term and **woody biomass** due to the nature of biomass sources. For example, cassava, cotton and coffee are all agricultural crops but their stems are wood. For convenience, non-woody biomass includes mostly agricultural crops, shrubs and herbaceous plants. See also **woody biomass**

Normalized vegetation index (NVI)   Refers to the ratio of the red to near-infrared reflection from vegetation as detected by remote sensing

Oven wood   Wood that has been split into fairly short, thin pieces and then air-dried

Oven-dried   Wood dried at constant weight in a ventilated oven at a temperature above the boiling point of water, generally 103–102 °C. It means that a fuel has zero moisture content and is sometimes referred to as *bone dry*

Oven-dried weight   The weight of a fuel or biomass material with zero moisture content

Photosynthesis   A term used commonly to denote the process by which plants synthesize organic compounds from inorganic raw materials in the presence of sunlight. All forms of life in the universe require energy for growth and maintenance. It is therefore the process whereby green plants use the sun's energy to produce energy-rich compounds, which may then be used to fix carbon dioxide, nitrogen and sulphur for the synthesis of organic material

Pol   (short for polarization) An analytical method to determine the level of sucrose in sugar cane. It is the apparent sucrose content of any substance expressed as a percentage by mass; it is often cited as a mixed juice pol, with world average value around 12 or 13. See also **Brix**

Primary energy (or primary production)   This measures the potential energy content of the fuel at the time of initial harvest, production or discovery prior to any type of conversion. It is often used for recording the total energy consumption of a country, which is misleading because it ignores the conversion efficiencies at which the fuel is used

Primary productivity   The next yield of dry plant matter produced by photosynthesis within a given area and period of time. Photosynthesis efficiency is the main determining factor in primary productivity. The annual global primary productivity is from $100–125 \times 10/9$ tonnes on land plus $44–55 \times 10/9$ tonnes in the world's oceans. Most of this biomass (44.3 per cent) is formed in forests and woodlands

Random sampling   In a random sampling, the choice of each sampling unit selected for measurement is made independently of that of any others; that is, the selection of any one unit gives no indication of the identity of any other selected unit. Selection must embody the Laws of Change so that each unit in the sampling frame has a known possibility of selection

Remote sensing   Surveying the earth's surface from aircraft or satellites using instruments to record different parts of the electromagnetic spectrum; it is also used to measure total biomass productivity

Renewable resources   Natural resources produced by photosynthesis, or derived from products of photosynthesis (e.g. energy from plants), or directly from the sun (e.g. solar energy) utilized by humans in the form of plant or animal products. See **renewable energy**

Renewable energy   Refers to an energy form the supply of which is partly or wholly generated in the course of the annual solar cycle. The term covers those continuous flows that occur naturally and repeatedly in the environment (e.g. energy from the sun, the wind, from plants, etc.). Geothermal energy is also usually regarded as a renewable energy source since, in total, it is a resource on a vast scale

Secondary energy (or secondary fuels or final energy)   Differs from **primary energy** by the amount of energy used and lost in supply-side conversion systems, that is, sources of energy manufactured from basic fuels (e.g. in biomass gasifiers, charcoal kilns)

Site class   A measure of the relative productive capacity of a site for the crop or stand under study, based on volume or weight or the maximum mean annual increment that is attained or attainable at a given age

Site index   See **site class**

Solid volume (wood)   Volume of the actual logs only, established by taking individual dimensions of evenly cut geometries of these logs; usually measured in cubic metres

Stumpage   The value of a standing tree. The terms *stumpage fee, stumpage value, royalty* and *stumpage tax* are used interchangeably, often causing confusion. Stumpage fee is the financial concept of the value of standing wood resources. Stumpage value is the economic concept of the worth to society of a unit of wood resource, estimated using the economic concepts of real resource (opportunity) costs and shadow (efficiency) prices. Stumpage royalty (royalty), presently refers to payments extracted by a public authority (e.g. government) in exchange for use of a tree on public land

Stacked volume (wood)   Volume occupied by logs when they are stacked closely to form heaps of given dimensions. Usually measured in cubic metres

Stand   A continuous group of trees sufficiently uniform in species composition, arrangement of age classes and conditions to be a homogeneous and distinguishable unit

Standing volume   The volume of standing trees, all species, living or dead, all diameters down to a stated minimum diameter. Species which do not have an upright trunk (brush, etc.) are not considered as trees. It includes dead trees lying on the ground which can still be used for fibre or as fuel

Standing volume tables   Instead of felling sample trees or measuring them standing or using a single tree volume table, volume per ha of even-aged crops of a single species may be predicted directly using a stand volume table. The commonest stand volume table is derived from simple linear regression of volume per ha, V, or on the combined variable-basal area per ha multiplied by some measure of height representative of the crop; often dominant height is used because it is convenient, being objectively defined as the height of the 100 fattest or tallest trees per ha. Stand volume tables normally have confidence limits at 0.05 of about ±5–10 per cent of the mean stand volume using 20 samples per stand

Stere   A measure of stacked wood. By definition a stere is a stack of wood 1 m long by 1 m wide by 1 m high having a total volume of 1 m^3. In practice, the weight of a stere ranges between about 250 to 600 kg

Stock   The total weight of biomass at any given time. Preferably, biomass stock or inventory should be expressed initially as total gross figures, and then as oven-dried weight

Stock energy units   These measure a quantity of energy in a resource or stock (e.g. wood energy in a tree at a given point in time); examples are tons oil equivalent or multiples of the joule (MJ, GJ, PJ)

Stocking (wood)   The volume or weight of woody material on unit area, often expressed as a proportion of a measured or theoretical maximum possible under the particular soil and climatic conditions prevailing. Often the proportion of the unit area covered by tree crowns is used as a rough guide to stocking

Total biomass volume   All above-ground parts of a plant

Tree height measurement   Tree height measurement is important as it is often one of the variables commonly used in the estimation of tree volume. Tree height can have different meanings which can lead to practical problems. Thus, it is important that the terms are clearly defined:

*   *Total height* refers to the distance along the axis of the tree stem between the ground and the tip of the tree.
*   *Bole height* is the distance along the axis of the tree stem between the ground and the crown point.
*   *Merchantable height* refers to the distance along the axis of the tree stem between the ground and the terminal position of the last usable portion of the tree stem (minimum stem diameter).
*   *Stump height* is the distance between the ground and the basal position on the main stem where a tree is cut – about 30 cm.
*   *Merchantable length* refers to the distance along the axis of the tree stem between the top of the stump and the terminal position of the last usable portion of the tree stem.
*   *Defective length* is the sum of the portions of the merchantable length that cannot be utilized due to defects.
*   *Sound merchantable length* equals the merchantable length minus the defective length.
*   *Crown length* refers to the distance on the axis of the tree stem between the crown point and the tip of the tree.

Vegetation indices   These indices are derived from combined visible and near-infrared observations from terrestrial remote sensing. Vegetation indices are an important aspect of electromagnetic remote sensing which may be of great value to bioclimatic research

Woodfuels   All types of biofuels derived directly and indirectly from trees and shrubs grown in forest and non-forest lands. Woodfuels also include biomass derived from silvicultural activities (thinning, pruning, etc.) and harvesting and logging (tops, roots, branches, etc.), as well as industrial by-products derived from primary and secondary forest industries that are used as fuel. They also include woodfuel derived from ad hoc forest energy plantations. Woodfuels are composed of four main types of commodities:

- fuelwood (or firewood)
- charcoal
- black liquor and
- other.

According to origin, woodfuels can be divided into three groups as detailed below.

- Direct woodfuels – these consist of wood directly removed from:
    — forests (natural forests and plantations) – land with tree crown cover of more than 10 per cent and area of more than 0.5 ha);
    — other wooded lands (land either with a tree crown cover of 5–10 per cent of trees able to reach a height of at least 5 m at maturity in situ; or crown cover of more than 10 per cent of trees not able to reach a height of 5 m at maturity in situ, and shrub or bush cover);
    — other lands to supply energy demands, including both inventoried (recorded in official statistics) and non-inventoried woodfuels.
- Indirect woodfuels – these usually consist of industrial by-products, derived from primary (sawmills, particle board, pulp and paper mills) and secondary (joinery, carpentry) wood industries, such as sawmill rejects, slabs, edging and trimmings, sawdust, shavings and chips bark, black liquor, etc.
- Recovered woodfuels – refers to woody biomass derived from all economic and social activities outside the forest sector, usually wastes from construction sites, demolition of buildings, pallets, wooden containers and boxes, etc. (FAO definition)

Woody biomass   Comprises mainly trees and forest residues (excluding leaves), although some shrubs and bushes and agricultural crops such as cassava, cotton and coffee stems are wood and are often included. Woody biomass is the most important form of biomass energy. See also **biomass** and **non-woody biomass**

Yield   For plant matter yield is defined as the increase in biomass over a given time and for a specific area, and must include all biomass removed from the area. The yield or annual increment of biomass is expressed in dry tonnes/ha per year. It also should be clearly stated whether the yield is the current or mean annual increment

Most Commonly Used Biomass Symbols

Biomass and other

od or OD   = oven dry
odt or ODT = oven dry ton
ad or AD   = air dry
nc or MC = moisture content
mcwb   = moisture content, wet basis
mcdb   = moisture content, dry basis
MAI   = Mean Annual Increment
GHV   = Gross Heating Value
HHV   = Higher Heating Value
LHV   = Lower Heating Value
NHV   = Net Heating Value

Metric unit prefixes

Prefix	Symbol	Factor
exa	E	10^{18}
peta	P	10^{15}
tera	T	10^{12}
giga	G	10^{9}
mega	M	10^{6}
kilo	k	10^{3}
nano	n	$10{-}9$
micro	μ	$10{-}6$
milli	m	$10{-}3$

Metric equivalents

1 km = 1000 m
1 m   = 100 cm
1 cm = 10 mm
1 km^2 = 100 ha
1 ha   = 10,000 m^2

Costs

Multiply	by	To obtain
$/ton	1.1023	$/Mg
$Mg	0.9072	$/ton
$/MBtu	0.9470	$/GJ
$GJ	1.0559	$/MBtu

Energy Units: Basic Definitions

Energy can be measured in different units, among which the most generally used is **joule** (symbol: J). In the past, **calorie** (symbol: cal) was commonly used to measure heat (thermal energy). However, the 9th General Conference on Weights and Measures in 1948 adopted the joule as the unit of measurement for heat energy.

The energy content of **commercial energy sources** is expressed by the **calorific values**, which give the energy content of the given material related to the unit of weight, measured in **kJ/kg** (formerly in kcal/kg). The calorific values of various energy sources are always approximate figures since their precise values greatly depend on their chemical composition.

The quantities of **liquid fuels** and some solid fuels, like fuelwood, wood chips, etc. in many cases are measured by their volume and, consequently, their calorific values can be expressed in relation to their volumes as well. With the help of the **specific gravity** (measured in kg/litre) or the **bulk density** (measured in kg/m³) of the energy carriers, their calorific values, can be estimated. Calorific values can change greatly according to their chemical composition and physical state.

Electric energy: the energy of **electricity** is measured generally in **kilowatt-hours** (symbol: kWh); it is 3.6 million **J** (3.6 MJ), since 1 **J** = 1 watt-second which should be multiplied by 3600 (1 hour = 3600 s) and 1 k = 1000. On the other hand, 1 kWh of electric energy is equivalent to 860 kcal of heat, therefore the conversion factor from the old energy unit **calorie** to the standard unit of **joule** can be derived, as follows:

$$1 \text{ kcal} = 3600/860 = 4.1868 \text{ kJ}$$

The energy equivalent of 1 kWh = 3.6 MJ = 860 kcal relates to the **net (useful) energy** of electricity. However, due to the approximately 30 per cent efficiency of generation and transport of electricity energy, some 12,000 kJ equivalent to 2870 kcal of primary energy sources should be utilized for the generation of 1 kWh of electricity. Hence, when the energy equivalent of electricity is discussed, these two approaches should be clearly distinct.

Attention should be drawn to the fact that the **kilowatt-hour** is not a unit of power but a unit of energy, and it is equal to the total quantity of energy dissipated when a power of 1 kW operates for a period of 1 hour.

Energy equivalent and conversion factors

In order to compare the energy equivalent of various energy sources and energy carriers having different calorific values and physical-chemical characteristics, various **energy units** and **energy equivalents** are used in the technical-economic calculations.

Although the basic units are **joule** and **kWh**, in practice, however, these are very small and units and standard prefixes as detailed in the metric unit prefixes table in Appendix II.

Prefix	Symbol	Factor
exa	E	10^{18}
peta	P	10^{15}
tera	T	10^{12}
giga	G	10^{9}
mega	M	10^{6}
kilo	k	10^{3}

On the other hand, **kWh** relates first to electrical energy, and is not a very convenient unit for measuring and expressing the energy content of fossil fuels and solid fuels. Previously, the **kg of coal equivalent** (kgCE; 30 MJ/kg LHV) was common but, nowadays, the **kg of oil equivalent** (kgOE; 42 MJ/kg LHV) is usually the generally accepted unit for the expression of the calorific equivalent of different energy carriers.

Power and efficiency

The **mechanical power** of internal combustion engines was previously measured in **HP** (horse power), but the new standard unit is **kW** (conversion factor is HP = 0.736 kW).

The **thermal power** of a heat generator was previously expressed in **kcal/h**; the new standard unit is also **kW** (the conversion factor is 1 kW = 860 kcal/h so, for example, the thermal power of a 100,000 kcal/h boiler is 100,000/860 = 116 kW.

The **electric power** is still measured in **kW** (being the fraction of a **joule** per second – J/s).

All energy sources and various forms of energy can be transformed to any other forms, but the energy transformation processes are always accompanied by larger or smaller amounts of energy losses, therefore, the **efficiency of energy transformation** largely depends on the type of energy source or carriers, the construction and design of the energy transformation equipment and the actual operation conditions. Therefore, when the various combustion units and heating systems are compared from energetic and economic points of view, apart from the **specific fuel price** of the various energy carriers expressed in **currency unit per kJ** or **kgOE** (e.g. US$/kJ or US$/kgOE), the **specific net (usable) energy price**, taking into consideration the expected energy efficiency should be calculated.

International units

J = Joule
1 ha = hectare = 2.47 acres
t = metric tonne = 1,000 kg.
1 btu (British Thermal Unit) = 1.054 kJ
1 calorie = 4.19 J
1 kWh = 3,600 J
1 W = 1 J s − 1

Further reading

Bialy, J. (1979) *Measurement of Energy Released in the Combustion of Fuels*, School of Engineering Sciences, Edinburgh University, Edinburgh
Bialy, J. (1986) *A New Approach to Domestic Fuelwood Conservation: Guidelines for Research*, FAO, Rome
www.convertit.com (electronic unit converter)
www.exe.ac.uk/trol/dictunit/ (a general dictionary of units)

Some Conversion Figures of Wood, Fuelwood and Charcoal

Table IV.1 *Conversion figures[a] (air dry, 20 per cent moisture)*

1 t wood	=	1000 kg
	=	1.38 m^3
	=	0.343 t oil
	=	7.33 barrels oil
	=	90.5 litres kerosene
	=	3.5 Mkcal
1 t wood (db)	=	15 GJ
1 t wood (od)	=	20 GJ
1 m^3 wood	=	0.725 t
	=	30 lb
1 m^3 wood (stacked)	=	0.276 cord (stacked)
1 cord[b] wood	=	1.25 t (3.62 m^3) (general)
1 cord (stacked)	=	2.12 m^3 (solid)
	=	3.62 m^3 (stacked)
	=	128 ft^3 (stacked)
1 stere[b] wood	=	1 m^3 (0.725 t)
1 pile wood	=	0.510 t (510 kg)
1 QUAD	=	62.5 t wood (od)
	=	96.2 t wood (green)
1 headload[c]	=	37 kg
1 t charcoal	=	6–12 t wood[d]; 30 GJ
1 m^3 charcoal	=	8.28–16.56 m^3 wood
	=	0.250 t
1 kg charcoal	=	4.2–4.7 m^3 producer gas
1 kg wet wood	=	1.2–1.5 m^3 producer gas
1 kg dry wood	=	1.9–2.2 m^3 producer gas
1 t agricultural residues	=	10–17 GJ[e]

Notes:

[a] These are approximate figures only and therefore it is important that all conversion factors used are clearly stated.

[b] Cord and stere measures can vary appreciably in actual practice. See Appendix I Glossary of Terms.

[c] The headload varies from place to place. This corresponds to a mature female.

[d] This wide variation is due a number of factors (e.g. species, moisture content, wood density, charcoal piece size, fines, etc.).

[e] This large variation is due to moisture content. With a moisture content-s of 20 per cent it is 13–15 GJ.

Source: Rosillo-Calle et al (1996).

References

Rosillo-Calle, F., Furtado, P., Rezende, M. E. A. and Hall, D. O. (1996) *The Charcoal Dilemma: Finding Sustainable Solutions for Brazilian Industry*, Intermediate Technology Publications, London

More on Heating Values and Moisture Contents of Biomass

Appendix 2.3 gives a detailed account on how to calculate volume, density, moisture content and energy values of biomass. Appendix 5 provides additional details on energy values and mositure content.

To recall, the heating value (HV) of fuels is recorded using two different types of energy content: **gross** and **net**. Although for petroleum the difference between the two is rarely more than about 10 per cent, for biomass fuels with widely varying moisture contents the difference can be very large.

Gross Heating Value (GHV)

This refers to the total energy that would be released through combustion divided by the weight of the fuel. It is also called Higher Heating Value (HHV).

Net Heating Value (NHV)

This refers to the energy that is actually available from combustion after allowing for energy losses from free or combined water evaporation. The NHV is always less than GHV, mainly because it does not include two forms of heat energy released during combustion:

- the energy to vaporize water contained in the fuel, and
- the energy to form water from hydrogen contained in hydrocarbon molecules, and to vaporize it.

Furthermore the difference between NHV and GHV depends largely on the water (and hydrogen) content of the fuel. Petroleum fuels and natural gas contain little water (3–6 per cent or less) but biomass fuels may contain as much as 50–60 per cent water at point of combustion.

Heating values of biomass fuels are often given as the energy content per unit weight or volume at various stages: **green, air-dried** and **oven-dried** material (see Appendix 2.3 and Glossary of Terms).

Most international and national energy statistics are now given in terms of the LHV, in which one tonne of oil equivalent is defined as 1010 calories (107 kilocalories), or 41.868 GJ. However, some countries and many biomass energy reports and projects still use higher heating values. For fossil fuels such as coal, oil and natural gas, and for most forms of biomass, the lower heating value is close to 90 per cent of the higher heating value.

Moisture content

The water contained within biomass material can alter by a factor of 4–5 between initial harvesting (as 'green' crop or wood) and its use as a fuel after some time, during which the material partially dries or loses water. The water content at any stage in this process, usually termed moisture content and given as a percentage, is measured in two ways, on a 'wet' and a 'dry' basis (see also Appendix 2.3):

- moisture content, dry basis (mcdb) is the weight of water in the biomass divided by the dry weight of biomass;
- moisture content, wet basis (mcwb) is the weight of water in the biomass divided by the total weight of the biomass; that is, the weight of water plus the weight of dry biomass.

The relationship between mcdb and mcwb is illustrated in the first two columns of Table V.1 that gives the HHV and LHV of biomass, in GJ per tonne, according to moisture content and three typical values of the HHV of dry biomass (HHVd); see Kartha et al (2005) pp141–143 for further details.

As can be appreciated from Table V.1 moisture has a large effect on the energy content per unit mass of biomass, whereas the use of HHV has a smaller but significant effect. As can be expect, the difference between the lower and higher heating values increases with moisture content (Kartha et al (2005), pp141–143). Table V.2 illustrates residue product, energy value and moisture content.

Table V.1 *Energy content of biomass according to moisture content and use of higher versus lower heating values*

Moisture (%)		Higher heating value of dry biomass (HHVd)					
Wet basis	Dry basis	22 GJ	20 GJ	18 GJ	22 GJ	20 GJ	18 GJ
		Higher (HHV) and lower (LHV) heating values of biomass at given moisture content					
(mcwb)	(mcdb)	HHV	HHV	HHV	LHV	LHV	LHV
0	0	22.0	20.0	18.0	20.79	18.79	16.79
5	5	20.9	19.0	17.1	19.64	17.74	15.84
10	11	19.8	18.0	16.2	18.49	16.69	14.89
15	18	18.7	17.0	15.3	17.34	15.64	13.94
20	25	17.6	16.0	14.4	16.18	14.58	12.98
25	33	16.5	15.0	13.5	15.03	13.53	12.03
30	43	15.4	14.0	12.6	13.88	12.48	11.08
35	54	14.3	13.0	11.7	12.72	11.42	10.12
40	67	13.2	12.0	10.8	11.57	10.37	9.17
45	82	12.1	11.0	9.90	10.42	9.32	8.22
50	100	11.0	10.0	9.00	9.27	8.27	7.27

Source: See Kartha et al (2005).

Table V.2 *Residue product, energy value and moisture content*

Biomass item	Ratio of product: residue	Product energy value (GJt⁻¹)	Product moisture status	Residue energy value (GJt⁻¹)	Residue moisture status
Course grains[a]	1.0:1.3	14.7	20% air dry	13.9	20% air dry
Oats[a]	1.0:1.3	14.7	20% air dry	13.9	20% air dry
Maize[a]	1.0:1.4	14.7	20% air dry	13.0	20% air dry
Sorghum	1.0:1.4	14.7	20% air dry	13.0	20% air dry
Wheat	1.0:1.3	14.7	20% air dry	13.9	20% air dry
Barley	1.0:2.3	14	20% air dry	17.0	Dry weight
Rice	1.0:1.4	14.7	20% air dry	11.7	20% air dry
Sugar cane	1.0:1.6	5.3	48% moisture	7.7	50% moisture
Pulses total[b]	1.0:1.9	14.7	20% air dry	12.8	20% air dry
Dry beans	1.0:1.2	14.7	20% air dry	12.8	20% air dry
Cassava[c]	1.0:0.4	5.6	Harvest	13.1	20% air dry
Potatoes	1.0:0.4	3.6	50% moisture	5.5	60% moisture
Sweet potatoes[c]	1.0:0.4	5.6	Harvest	5.5	Harvest
Fruits	1.0:2.0	3.2	Harvest	13.1	20% air dry
Vegetables	1.0:0.4	3.2	Fresh weight	13.0	20% air dry
Fibre crops[d]	1.0:0.2	18.0	20% air dry	15.9	20% air dry
Seed cotton	1.0:2.1	25.0	Dry weight	25.0	Dry weight
Sunflower[e]	1.0:2.1	25.0	Dry weight	25.0	Dry weight
Soybeans	1.0:2.1	14.7	20% air dry	16.0	20% air dry
Groundnuts[e]	1.0:2.1	25.0	20% air dry	16.0	20% air dry
Tea	1.0:1.2	10.2	20% air dry	13.0	20% air dry
Copra (coconut product)		28.0	5% moisture		
Fibre (coconut residue)	1.0:1.1			16.0	Air dry
Shell (coconut residue)	1.0:0.86			20.0	Air dry

Notes:

[a] Values for course grains (including maize, sorghum and oats) are best assumptions based on those values for similar crops given by Ryan & Openshaw, 1991; Senelwa & Hall, 1994; Strehler & Stutzle, 1987; and Woods, 1990.

[b] Values for pulses total are best assumptions based on those values for similar crops given by Ryan & Openshaw, 1991; Senelwa & Hall, 1994; and Strehler & Stutzle, 1987.

[c] Values for cassava and sweet potatoes are best assumptions based on those values for similar crops given by Senelwa & Hall, 1994; and Strehler & Stutzle, 1987.

[d] Values for fibre crops are best assumptions based on those values for similar crops given by Senelwa & Hall, 1994; and Strehler & Stutzle, 1987.

[e] The product : residue ratios for sunflower and groundnuts are best assumptions based on those values for similar crops given by Strehler & Stutzle, 1987.

Source: Hemstock and Hall (1995); Hemstock (2005).

References and further reading

Hemstock, S. L. (2005) *Biomass Energy Potential in Tuvalu* (Alofa Tuvalu), Government of Tuvalu Report

Hemstock, S. L. and Hall, D. O. (1995) 'Biomass energy flows in Zimbabwe', *Biomass and Bioenergy*, 8, 151–173

Kartha, S., Leach, G. and Rjan, S. C. (2005) *Advancing Bioenergy for Sustainable Development; Guidelines for Policymakers and Investors*, Energy Sector Management Assistance Programme (ESMAP) Report 300/05, The World Bank, Washington, DC

Leach, G. and Gowen, M. (1987) *Household Energy Handbook: An Interim Guide and Reference Manual*, World Bank Technical Paper no 67, World Bank, Washington DC

Openshaw, K. (1983) 'Measuring fuelwood and charcoal', in *Wood Fuel Surveys*, FAO, pp173–178

Ryan, P. and Openshaw, K. (1991) *Assessment of Bioenergy Resources: a discussion of its needs and methodology.* Industry and Energy Development Working Paper, Energy Series Paper no 48. World Bank, Washington DC, USA

Senelwa, K. A. and Hall, D. O. (1994) 'A biomass energy flow chart for Kenya', *Biomass and Bioenergy*, 4, 35–48.

Strehler, A. and Stutzle, G. (1987) *Biomass Residues.* D. O. Hall and R. P. Overend (eds), Biomass: Regenerable Energy, Elsevier Applied Science Publishers, London, UK

Woods, J. (1990) *Biomass Energy Flow Chart for Zambia – 1988.* King's College London, Division of Biosphere Sciences, London, UK

www.eere.energy.gov/biomass/feedstock_databases.html (conversion factors)

www.fao.org//doccrep/007/ (bioenergy terminology, parameters, units and conversion factors, properties of biofuels (moisture content, energy content, mass, volume and density)

Measuring Sugar and Ethanol Yields

Sugar yields

On average, for every 100 t of sugar cane ground the following is produced (Thomas, 1985):

- 1.2 t of raw sugar (98.5 pol);
- 5.0 t of surplus bagasse (49 per cent moisture) = 1300 kWh surplus electricity (depending on the boiler efficiency);
- 2.7 t of molasses (89 Brix; specific gravity 1.47);
- 3.0 t of filter mud (80 per cent moisture);
- 0.3 t of furnace ash;
- 30.0 t of cane top/trash (barbojo).

These quantities are industrial averages and, therefore, vary considerably between countries and between installations within the same country.

Ethanol yields

1 kg invert sugar	=	0.484 kg ethanol	=	0.61 litre
1 kg sucrose	=	0.510 kg ethanol	=	0.65 litre
1 kg starch	=	0.530 kg ethanol	=	0.68 litre

1 kg ethanol	=	6,390 kcal
1 litre ethanol	=	5,048 kcal
1 ton ethanol	=	1,262 litres
	=	26.7 GJ (21.1 MJ/l), LHV; 23.4 GJ/l, HHV

Note: Maximum theoretical yield of ethanol on a weight basis is 48 per cent of converting glucose.

Source: Biomass Users Network (BUN) figures; http://bioenergy.ornl.gov/papers/misc/energy_conv.html

Properties of sugar cane

In general, 1 tonne of sugar cane (UNICA, Brazil – www.portalunica.com.br):

- has an energy content of 1.2 boe;
- can produce 118 kg of sugar (+ 10 litres of molasses);
- can produce 85 litres of anhydrous ethanol;
- can produce 89 litres of hydrated ethanol.

Conversion factors: litre ethanol to litre of gasoline

The conversion factor is not simple because it is largely dependent not only on the calorific value of the fuels (i.e. ethanol and gasoline) but on the engine efficiency, tuning, blending, etc. For example, while the calorific value of ethanol is much lower, when it burns it does so with much greater efficiency than gasoline, thus producing more power, partially offsetting its lower calorific value. In Brazil, for example, where ethanol is blended with gasoline in a proportion of 24 per cent,[1] the conversion factor varies between 1.2 to 1.3 litres ethanol/litre gasoline, to take into account the lower heating value and the higher efficiency of ethanol in engines.

This conversion factor is even more complicated in the case of the flex-fuel vehicles because the blending proportion can vary considerably (i.e. in Brazil from a minimum of 20 per cent to almost 100 per cent blend, although 50–50 per cent ethanol–gasoline blend is currently common). But the complication does not end here because the engine characteristics of the flex-fuel vehicles are different as the compression ratio of these engines varies, for example, currently this is between gasoline and ethanol fuel engine (see Rosillo-Calle and Walter, 2006 for further details). The flex-fuel vehicle engine has a compression ratio somehow between gasoline and ethanol e.g. the compression ration of the VW Golf 1.6 is *10.1* and that of Fiat Palio 1.3 8v is *11.1*. (Note: both cars are total flex-fuel vehicles; the compression ratio is being improved and it is possible that in near future new technological development will allow this compression ratio to improve even further)

Sugar and ethanol production from sugar cane

The flow diagram in Figure VI.1 illustrates in simplified form the production of sugar and ethanol from sugar cane in Brazil.

Ethanol agro-industrial processes

Table VI.1 details the energy flows involved in ethanol agro-industrial processes in Brazil.

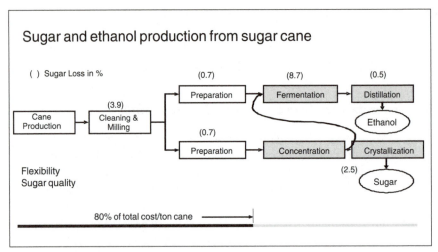

Source: Macedo (2003).

Figure VI.1 *Sugar and ethanol production from sugar cane in Brazil*

Table VI.1 *Energy balance of ethanol agro-industrial processes in São Paulo, Brazil*

| | Energy flows (MJ/ tonne of sugar cane) | | | |
| | Average | | Best case | |
	Consumption	Production	Consumption	Production
Agricultural activities	202		192	
Industrial process	49		40	
Ethanol production		1922		2052
Bagasse surplus		169		316
Total	251	2090	232	2368
Energy output/input	8.3		10.2	

Source: Macedo et al (2004).

Note

1   This can vary, but in most cases it stays between 20–25 per cent blend.

References

Macedo, I. C. (2003) 'Technology: Key to sustainability and profitability – A Brazilian view', Cebu, Workshop Technology and Profitability in Sugar Production, International Sugar Council, Philippines, 27 May 2003

Macedo, I. C., Lima Verde Leal, R. and Silva, J. E. A. R. (2004) 'Assessment of greenhouse gas emission in the production and use of fuel ethanol in Brazil', Secretariat of the Environment, Government of the State of São Paulo, SP

Rosillo-Calle, F. and Walter, A. (2006) 'Global market for bioethanol: Historical trends and future prospects', *Energy for Sustainable Development* (Special Issue), March

Thomas, C. Y. (1985) *Sugar: Threat or Challenge?*, International Development Research Centre, IDRC-244e, Ottawa, Canada

www.portalunica.com.br (information on Brazil's sugar cane energy)

Carbon Content of Fossil Fuels and Bioenergy Feedstocks

Carbon content of fossil fuels

1 tonne of coal	=	750 kg
1 ton of oil	=	820 kg
1 tonne of natural gas	=	710 kg
1 m^3 natural gas	=	0.50 kg
1 litre oil (average)	=	0.64 kg
1 litre diesel	=	0.73 kg
1 tonne diesel fuel	=	850 kg[a]
1 tonne petrol/gasoline	=	860 kg[b]

Carbon content of oven dry bioenergy feedstocks (approximate)

Woody crops	=	50%
Graminaceous (grass and agricultural residues)	=	45%

Notes:

[a] 1159 litres per tonne at 30 °C.
[b] 1418 litres per tonne at 30 °C.

Adapted from: http://bioenergy.ornl.gov/papers/misc/energy_conv.html.

Index

Note: page numbers in *italics* denote references to Figures and Tables; page numbers in **bold** denote references to Appendices.